HANDBOOK TO IDAHO'S SAWTOOTH COUNTRY FLORA

HANDBOOK TO IDAHO'S SAWTOOTH COUNTRY FLORA

Sawtooth Mountains · White Cloiud Mountains · Boulder Mountains · Smoky Mountains · Pioneer Mountains · Samon River Mountains

by
Ray S. Vizgirdas
Edna M. Rey-Vizgirdas

Copyright © 2018 by Ray S. Vizgirdas

Photographs by Ray S. Vizgirdas
Illustrations by Edna M. Rey-Vizgirdas

All rights reserved. This book or any portion thereof may not be reproduced or used in any manner whatsoever without the express written permission of the publisher except for the use of brief quotations in a book review or scholarly journal.

First Printing: 2017

ISBN # 978-1-387-79532-1

Mountainforaging@gmail.com

Disclaimer

While this book documents the uses of wild plants within Idaho's Sawtooth Country, the authors disclaimsany liability for injury that may result from following any instructions for collecting, preparing or consuming plants described in this guide. Efforts have been taken to assure the descriptions of plants represented are accurate representations of the family, genus, and species noted. It should be understood that growth conditions, improper identification, and varietal differences, as well as an individual's own sensitivity or allergic response can contribute to a hazard in sampling or using a plant. Furthermore, the reader is encouraged to seek the assistance from experienced botanists in identifying any of the plants discussed in this book.

Acknowledgements

This handbook is the result of mor than 12 years working on the Sawtooth National Forest (SNF). As a former U.S. Fish and Wildlife Service biologist (RSV) assigned to work with the SNF on endangered species issues, I got to know the landscape very well. I am indebted to several individuals for sharing their knowledge – I hope I was a good student. Thank you to Deb Bumpus, Tom Bandolin, Deb Carter, Johnna Roy, Robin Garwood, Mark Moulton, Deb Taylor, Christne Gertschen, and Terry Clark. We had fun and we made a difference.

INTRODUCTION

The aim of this handbook is to provide a guide to the common and interesting vascular plants found in Idaho's Sawtooth Country. We provide information on the natural history, ecology, and adaptations of common trees, shrubs, and wildflowers, as well as how the plants were used by Native Americans and early inhabitants of the area.

The Idaho Sawtooth Country (hereafter referred to as **Sawtooth Country**) is loosely described as the area that lies in the heart of the approximately 2-million-acre Sawtooth National Forest in central Idaho. Within this area are the Sawtooth Wilderness, Sawtooth National Recreation Area, and the recently designated wilderness areas collectively known as the "Boulder-White Clouds." The Boulder-White Clouds is made up of three distinct wilderness regions: the Hemingway-Boulders, the White Clouds, and the Jim McClure-Jerry Peak.

The Sawtooth Country is located on the northern edge of the Great Basin Desert and as such numerous sagebrush steppe species of plants make their way into and around the four mountain ranges of the area. The mountain ranges include: Sawtooth Range, Boulder Mountains, White Cloud Mountains, and Smoky Mountains. Portions of the Pioneer and Salmon River Mountains are also covered in this handbook.

Sawtooth Range

The Sawtooth Range is an oval-shaped group of mountains is bounded by the Sawtooth Valley on the east and by Idaho State Highway 75 in the north and west. The Boise and Smoky mountains comprise the southern boundary of the Sawtooth's, which ends on the slopes of Greylock Mountain just north of the town of Atlanta. Scattered throughout the range are numerous peaks that exceed 10,000 feet. The Sawtooth crest runs for about 32 miles from north to south and measures about 20 miles across at its widest point.

Geologically, the Sawtooth's are made up of granite and glaciation has had a significant role in shaping the mountains you see today. The Sawtooth granite is extensively fractured by vertical jointing and this has caused the range to erode into the jagged edges and towers you see, rather than the hump-shaped appearance that is characteristic of the Idaho Batholic mountains elsewhere (e.g., the Salmon Mountains to the north). During glaciation, millions of tons of rock where cut away to create to create the state's largest alpine lake basins.

Boulder Mountains

These mountains are best known for their multi-colored southern escarpment which rises abruptly along Idaho State Highway 75 just north of the town of Ketchum. The range extends for about 50 miles - from Ketchum to Challis - and is flanked on the west by Idaho State Highway 75 between Ketchum and Galena Summit, the White Cloud mountains and East Fork Salmon, and in the north by the main Salmon River. The eastern boundary is U.S. Highway 93 between Challis and Thousand Springs Valley, and the Pioneer Mountains make up the southern boundary.

The Boulder's are comprised of granite, Challis volcanics, and highly metamorphosed rocks that have been extensively shattered and faulted. The southern front of the Boulder's rise above the Big Wood River Valley floor, is comprised of three layers of material. The top layer is made up of andesite and dacite, which were deposited by the Challis volcanics about 60 million years ago. The middle layer is pink granite which was intruded about 50 million years ago, and the bottom layer is the oldest and is composed of light-colored limestones and quartzites.

White Cloud Mountains

The White Cloud Mountains are approximately 20-miles long and 16-miles wide. The Salmon River flanks its western and northern boundaries, while the East Fork Salmon flanks the range on its eastern side. To the south lie the Boulder Mountains and because there is no definite division between the two, you often see the Boulder and White Clouds referred to as the "Boulder-White Clouds."

There are also extensive deposits of white limestone in the northwestern portion of the range, which when first viewed do look like white cumulus clouds (hence the name White Clouds). The highest peak is Castle Peak at 11,815 feet, and the well-known group of 14 or so "White Cloud Peaks" are all over 10,000 feet.

Smoky Mountains

The Smoky Mountains are located on the west side of the Big Wood River Valley, where the well-known towns of Ketchum and Sun Valley occur. The Smoky extends south from the Sawtooth Range 40 miles to the flat Camas Prairie. The Big Wood River and Idaho State

Highway 75 form the eastern boundary of the range, while the South Fork Boise River and the Soldier Mountains flank the range on its western side.

The peaks are composed of deteriorated metamorphic and granite-like rocks that reach a high point of 10,441 feet on Saviers Peak. More recently, block faulting has occurred which caused the Wood River Valley to sink in relation to the Smoky and Boulder mountains. As such, the Smoky mountains have a steep eastern front that contrast sharply with the range's western slopes, which merge almost invisibly into the Boise and Soldier mountains.

Pioneer Mountains

The Pionners are a popular hiking area for the residents of Sun Valley and Kethcum. The Pionners are described as compact, only about 12 by 20 miles in area. There are a number of alpine summits, including Hyndman's Peak which is about 12,009 feet.

Salmon River Mountains

Lastly are the mountains north of Stanley. Further north these mountains become part of the Frank Church-River of No Return Wilderness. This book has application to a major part of this wilderness area as well.

Sawtooth Country Plants

The common and easily identifiable vascular plants (ferns, gymnosperms, and flowering plants) are included in this handbook. No attempt was made to include every species that have been used in the past by Native Americans as some of them are just too dangerous to experiment with or are not worth the effort (calorie-wise). Also, not included are the other "plants" such as mushrooms, mosses, and lichens.

Imagine adding wild onions to your chicken ramen soup or nibbling twisted-stalk stems for its cool cucumber-taste. For the more adventurous, try making hot cider from kinnikinnick berries, you'll really appreciate the "wild" tastes out there.

Other tasty edibles with the Sawtooth Country include watercress (*Nasturtium*), spring beauty (*Claytonia*), twisted-stalk (*Streptopus*), mountain dandelion (*Agoseris*), lamb's quarters (*Chenopodium*), cheeseweed (*Malva*), clovers (*Trifolium*), stinging nettle (*Urtica*), and more! All this is free if you know how and where to look. This handbook will help.

Certainly, some of the plants within the area have medicinal use, however, I've kept this element to a minimum as not everyone who has an interest in wild foods shares the same interest for medicinal plants. If you are interested in medicinal plants of Idaho, I encourage you to seek out my other publications listed in the following reference section.

References The following references provide additional information about the plants found in the Sawtooth Country:

Vizgirdas, R.S. 2017a. *Field Guide to Food & Survival Plants along Idaho's Centennial Trail.* RSVizgirdas through Lulu.com

Vizgirdas, R.S. 2017b. *Wild Edible and Useful Plants of Idaho.* RSVizgirdas through Lulu.com

Vizgirdas, R.S. 2017c. *Field Guide to South-central Idaho Plants.* RSVizgirdas through Lulu.com

Vizgirdas, R.S. 2007. *A Guide to Plants of Yellowstone and Grand Teton National Parks.* University of Utah Press

Vizgirdas, R.S. 2003. *Useful Plants of Idaho.* Idaho State University Press.

DISTRIBUTION OF PLANTS IN THE SAWTOOTH COUNTRY

The vegetation within the Sawtooth Country is highly variable. Arid zones blend with higher elevation lands that receive abundant precipitation. The low productive drier sites that support sage-grass communities give way to mountain brush and tree communities. As elevation and moisture increase, the tree communities give way to a grass-forb alpine-type community.

Plant habitats are largely defined by physical factors in the environment - air temperature, precipitation, soil, humidity, are but a few of these factors. The variety and number of plants are usually greatest in areas where two or more habitats converge. These areas are called ecotones and it is along the boundary; for example, between forest and meadow, that the conditions are suitable for organisms typical of both habitats. This "edge" is usually richer in wildlife than either meadow or forest alone. There are numerous ecotones occurring in the Sawtooth Country.

The ecosystems within the Sawtooth Country are the result of continuous dynamic interactions between living organisms and their environment. Geologic events have changed the shape of the land, and shifting climates have converted forests into arid areas and back into forest again. In a sense, the landscape we see today is only a snapshot in the passage of time. In time, the ponds encountered during casual walks will become meadows, which eventually may be replaced by a forest. The colorful aspen groves will also gradually be succeeded by coniferous forests.

Following are descriptions of some common habitats found in the Sawtooth Country and the typical plants you can expect to find in them.

Lowland Riparian Areas

These types of ecosystems occur along the banks of rivers, streams, and other water bodies at the lower elevations. They also include floodplain woodlands and marshes that have various associations of herbs, shrubs, trees, and grasses that depend on a more or less continuous supply of water. These are narrow ecosystems that represent a transitional zone between the aquatic and terrestrial ecosystems, but usually have distinct vegetation and soils. In the southern portion of the area, large cottonwoods (*Populus*) and other flowering shrubs form thickets of varying width and density along the banks of streams. Some introduced species of plants have also escaped cultivation and have become established as common associates with the native cottonwoods.

Marshes occur in shallow margins of streams and ponds and elsewhere where standing water remains during most of the year. Species such as cattail and bulrush are conspicuous, and aquatic herbs such as pondweed, arrowhead, and water-plantain can also be found.

Sagebrush Shrubland

Sagebrush dominates extensive areas in the Sawtooth Country and the dense grayish-green stands vary in height from 2-6 feet tall. The dominant species is big sagebrush (*Artemisia tridentata*). Soils are usually deep, fine-grained and well-drained, and can be either acid or alkaline in nature.

Several species are associated with the mountain sagebrush, and often have a greater plant diversity than the adjacent coniferous forests. Rabbitbrush and bitterbrush are common associates, and wildflowers such as Mariposa lily, scarlet gilia, paintbrush, and lupine are also common in this sagebrush habitat.

Ponderosa Pine Forest

This forest type occurs on the west side of the Sawtooth Country. However, where Douglas-fir and ponderosa pine are common associates, ponderosa pines tend to occupy the warmest, driest forest sites.

Ponderosa pine stands tend to be open and parklike, with trees widely spaced to minimize competition for nutrients and water and the understory is often sparse. The amount of shrub and herbaceous understory is highly variable, depending on tree density and moisture availability. No understory plants are unique to ponderosa forests, but usually have species that are shared with neighboring communities.

Ponderosa pine is frequently found growing in mixed forests with aspen, Douglas-fir, and other tree species. The open, grassy ponderosa parklands support a greater variety of understory plants than to dense ponderosa stands.

Aspen Groves

Quaking Aspen (*Populus tremuloides*) is the dominant species of this habitat type. Aspen stands usually develop on recently burned sites, around the peripheries of meadows, forming narrow fringes, that are eventually invaded by conifers. They also occur on talus slopes. Aspen stands develop almost exclusively through root suckering.

A typical stand may consist of one or more clones. Trees are genetically identical. Aspen forests are the result of human or natural disturbance. Aspen ecosystems are important because they hold soils of disturbed areas in place and by doing so prevent stream siltation.

Open deciduous canopy allows understory vegetation to develop that are dominated by grasses, forbs, and well-developed shrub layers. Aspen ecosystems resemble mountain riparian systems in lushness of plant life.

Douglas-fir Forest

This is a climax conifer of the mid-slope forests, usually found above the Ponderosa and below the spruce-subalpine fir. In comparison to ponderosa pine forest, this forest type is distinctly cooler, shadier, and moister. Because little sunlight penetrates this forest, snows tend to last longer.

The species composition depends on local site conditions and history of disturbance and may vary considerably from one place to another. Douglas-fir is often mixed with other species such as ponderosa, aspen, Englemann spruce, lodgepole pine, and subalpine fir. At lower elevations plant species composition of Douglas-fir is similar to ponderosa, while at higher elevations, it may share many plant species associated with lodgepole and spruce-fir forests.

Lodgepole Pine Forest

Lodgepole Pine (*Pinus contorta*) is a wide-ranging North American conifer that very often forms pure stands in many areas. It is a pioneer species of burned ground and is highly intolerant of shade and depends on wildfires to create openings for regeneration. Unlike other trees that have built-in fire resistance, lodgepole pine has a thin, resinous, and highly flammable bark. The species can produce two types of cones: 1) serotinous, that require heat of a fire to open the cones, and; 2) non-serotinous, open irrespective of fire.

Lodgepole pine tend to occupy habitats that are too cold, too hot, too dry, or too wet for other conifers. In these areas they are the dominant tree.

Trees in this forest type are dense and usually even-sized, having the appearance of a tree plantation. The straight, pole-like appearance of their trunks gives the lodgepole pine its common name. In contrast to other forests (e.g., aspen community), lodgepole pine forests have very little understory.

Subalpine Forest

In the Sawtooth Country, subalpine fir and Engelmann spruce are the dominant species in the subalpine forest (often referred to as a spruce-fir forest). This forest type is essentially the southern extension of the boreal forest and these plants, in general, are better adapted to deep, long-lasting snowpacks, and a cool, brief growing season. The two species form closed forests on suitable sites in the lower subalpine and above the forest limit, they occur in scattered patches, islands, and ribbon forests. Up near timberline they form wind-flagged specimens and krummholz. The undergrowth of spruce-fir is sparse, usually mat forming huckleberries previal.

Mountain Riparian Systems

Deciduous, moisture loving trees and shrubs dominate this type of habitat. Riparian habitats are found throughout the area on moist sites but are not as extensive as most other types of mountain ecosystems. Most riparian systems are dominated by alder, cottonwood, willow, and birch. Structurally, they may consist of tree groves, shrub thickets, or a mixture of the two, sometimes interspersed with patches of wet meadow.

Tree groves and shrub thickets are usually found in the lower elevations, while in the higher elevations shrub thickets dominate. Riparian habitats of montane and subalpine regions have fewer species than at lower elevations, but diversity is still high compared to the adjacent upland ecosystems. The understory of riparian areas is usually lush and diverse when not disturbed. Forbs, grasses, sedges, rushes, mosses, lichens, and liverworts are present.

Mountain Meadows

Meadows are dominated by herbs as opposed to trees or shrubs. However, widely scattered trees and shrubs may be present. Some of the larger meadows are often referred to as "parks."

Alpine Tundra

This is the windswept, treeless area on the highest altitudes. Much of the alpine tundra appears to be barren rock, but many acres of tundra with deep soils have abundant small plants, insects, birds, and mammals. The tundra lies above the treeline. The plant communities vary tremendously in shape and plant composition, but these qualities are not conspicuously correlated with changes in slope exposure. Snow remains in addition to permanent snow fields. Plants species composition depends on local site characteristics, notably soil moisture, soil depth and rockiness, amount of snow accumulation, and exposure to wind.

Anthropogenic Habitats

Many places within the are have been altered by human activities. The original vegetative cover has been striped away and replaced by crop plants. Roads and housing developments have also been built. The removal of vegetative cover in an area opens it up for natural succession.

Some anthropogenic communities dominated by annuals, others by perennials. Some weed are grasses. Some weeds are adapted to one type of disturbance and not to another. Most weedy species produce large quantities of seeds and readily invade disturbed sites. Many have features that allow their seeds to be widely distributed. Some seeds remain dormant in the soil for many years, and when the soil is disturbed, the seeds germinate and establish a new successional community.

A Word on Weeds

Some of the plants covered in this book include species that are not native - variously called "weeds, invasives, and exotics." A few of these are descended from garden plants introduced in the early years of settlement in the area or having recently escaped from housing developments making their way into the native environments. Many have seeds that are easily carried in on the shoes and tires, within the fur of dogs and sheep, and by the wind. For the most part, these species are restricted in their distribution, growing in places where disturbance of the soil is routine, often following roadsides and trails. The species of greatest concern are those that have the ability to aggressively colonize and push aside native vegetation (e.g., cheatgrass [*Bromus tectorum*] and knapweed [*Centaurea*]).

Furthermore, a number of native species can also behave in a "weedy" manner. Often annuals, they have the ability to invade and maintain themselves on ground that is subjected to disturbances, especially at roadsides and along trails.

NUTRITION AND SEASONALITY OF PLANTS

Wild plants are a good source of vitamins and minerals. In fact, much of the medicinal value originally associated with many wild plants was due simply to their high vitamin and mineral content. Wild plants can also provide proteins, carbohydrates, fats, vitamins, and minerals that are needed for good mental and physical condition. Certain amino acids, however, can only be obtained from animal products (e.g., meats, milk, cheese, eggs). The following discussion summarizes some of the important vitamins and minerals that are needed for good health.

HOWEVER, while edible plants are common, they contain minimal food energy and take a long time to be assimilated by the body. The extra energy burned running around picking berries is usually higher than the caloric value of the food. For example, a person collecting plants on a cold rainy day for an hour or so will probably burn 400 calories. In order to come out even, that person would have to collect, clean, cook, eat, and digest more than two pounds of dandelion greens or 3 pounds of wild onions in an hour! If you going to be in the outdoors or on the trail for weeks or months, the vitamins and minerals contained in the plants would be useful as long as you gathered them on sunny warm days when issues like hypothermia are not a threat. In most cases, conserving energy and staying warm is much more important. People can survive weeks withot food as long as they have water.

Some Important Vitamins and Minerals for Health

Vitamins

These are organic compounds that are necessary in small quantities to prevent disease and help regulate the body's biochemical processes. Prolonged excessive doses of vitamins A, D, and K can have toxic effects. In addition to the vitamins listed below, biotin, choline, folic acid, and pantothenic acid are also essential nutrients.

Vitamin A - Vitamin A is not found in plants, but rather is manufactured by animals from pigments called carotenes which are common in plants. Vitamin A is essential for night vision, and promotes healthy skin and mucous membranes. It is also important for bones and teeth, proper digestion, and the production of red and white blood cells. It is fat soluble and sensitive to oxygen.

Vitamin B1 (Thiamine) - Found in both plant and animal tissues thiamine is important for the body's production of energy through the breakdown of carbohydrates. It appears to be important for normal functioning of nervous system and is involved in the action of the heart. Vitamin B1 is water-soluble and sensitive to heat. Most plants contain trace amounts.

Vitamin B2 (Riboflavin) - Riboflavin usually occurs in same foods as vitamin B1. It is essential for cell growth and enzymatic reactions by which the body metabolizes proteins, fats, and carbohydrates. Vitamin B2 is water-soluble and sensitive to light.

Vitamin B6 (Pyridoxine) - B6 is still a relatively little-known vitamin. It participates in many enzymatic reactions and is particularly important for brain and nervous system function. Vitamin B6 is water-soluble and sensitive to oxygen and ultraviolet light.

Vitamin B12 (Cyanocobalamin) - Little or no B12 is found in plants. Strict vegetarians sometimes suffer from pernicious anemia, a disease associated with B12 deficiency. Vitamin B12 is necessary for proper functioning of cells, especially in the nervous system, bone marrow, and gastrointestinal tract. It is involved in the metabolism of fats, proteins, and carbohydrates. B12 is water-soluble and sensitive to light, acids, and alkalis.

Vitamin C (Ascorbic Acid) - Vitamin C occurs in almost all plants to some degree. Since our bodies cannot make or store Vitamin C, a continuous supply must be present in the food we eat. Body cells require Vitamin C for proper functioning, as does the formation of healthy collagen (basic protein of connective tissue), bones, teeth, cartilage, skin, and blood vessels. Vitamin C also promotes the body's effective use of other nutrients such as iron, B vitamins, vitamins A and E, calcium, and certain amino acids. By promoting the formation of healthy connective tissue, Vitamin C helps to heal wounds and burns. Stress, fever, and infection tend to increase the body's need for Vitamin C. A deficiency of Vitamin C is called scurvy. Vitamin C is water-soluble and is sensitive to air, heat, light, alkalis, and copperware.

Vitamin D - Vitamin D does not occur in plants. However, some plants contain compounds called sterols, which when irradiated with ultraviolet light make Vitamin D. Vitamin D is necessary for healthy bones and teeth, and proper assimilation of calcium and phosphorus, and in preventing ricketts. It is a fat-soluble vitamin that is not sensitive to heat, light or oxygen.

Vitamin E (Tocopherol) - Vitamin E is found in both plant and animal tissue. It is an antioxidant, acting to protect red blood cells, Vitamin A, and unsaturated fatty acids from oxidation damage. It also helps maintain healthy membrane tissue. In laboratory experiments, it was found to be necessary for fertility in rats. Vitamin E is fat-soluble, and is sensitive to oxygen, alkali, and ultraviolet light.

Vitamin K - While Vitamin K occurs primarily in plants, it is also synthesized by intestinal bacteria found in the small intestine. It is necessary for the synthesis by the liver of the blood clotting enzyme prothrombin. Vitamin K is fat-soluble and is sensitive to light, oxygen, strong acids, and alcoholic alkalis.

Niacin (Nicotinic Acid) - A vitamin of the B complex, niacin occurs in both plant and animal tissue in various forms. In the body, niacin from plants is changed to niacinamide for use. Niacin takes part in enzyme reactions involved in the production of body energy and tissue respiration. Pellagra is a niacin deficiency disease. Niacin is water-soluble and is not sensitive to heat, acids, or alkali.

Minerals

These are chemical elements necessary for proper functioning of the body. Most are obtained from the foods we eat. There are two groups of minerals: macrominerals and microminerals. Macrominerals are found in relatively large amounts in the body, whereas microminerals are found in smaller amounts. Following is a list of minerals known to be necessary in human nutrition. There are other minerals, but their functions are not clearly understood.

Macrominerals

Calcium -This is the most abundant mineral in the body. It occurs in plants, dairy products, and seafood. Calcium is necessary for healthy bones and teeth, for clotting of blood, for the functioning of nerve tissue and muscles (including the heart), for enzymatic processes, and for controlling movement of fluids through cell walls.

Chlorine As a gas chlorine is poisonous, but in the form of chloride compounds, it is an essential mineral. It acts with sodium to maintain the balance between fluids inside and outside the cells. Gastric juices in the stomach contain hydrochloric acid, the production of which requires chloride. Table salt (NaCl) is our main source.

Magnesium - Found in both plant and animal tissue. Magnessium is essential as an enzyme activator and is probably involved in the formation and maintenance of body protein.

Phosphorus - Occurs in plant and animal tissue. Phosphorus takes part in the production of energy for the body, and is second only to calcium as a constituent of bones and teeth. Phosphorus is necessary for metabolic functions relating to the brain and nerves, as well as for muscle action and enzyme formation.

Potassium - Potassium is abundant in plant and animal tissue. It promotes certain enzyme reactions in the body, and acts with sodium to maintain normal pH levels and balance between fluids inside and outside the cells.

Sodium - This is a common mineral in plants and animals. It regulates the volume of body fluids and balanced with potassium, it helps maintain cell fluid equilibrium. It is also necessary for nerve and muscle functioning. The ideal amount can be obtained through a diet of vegetables such as dandelion greens, spinach, mustard greens, watercress, and carrots.

Sulfur - Sulfur supplies come from sulfur-containing amino acids and from the B vitamins thiamine and biotin. Main sources are dairy products, meats, legumes, nuts, and grains. Sulfur is involved in bone growth, blood clotting, and muscle metabolism. It also helps to counteract toxic substances in the body by combining with them to form harmless compounds.

Microminerals

Copper - Found in plant and animal tissue. Copper is essential (along with iron) for the formation of hemoglobin in red blood cells. Copper is also important for protein and enzyme formation, as well as for the nervous and reproductive systems, bones, hair, and pigmentation.

Iodine - The only dependable source of iodine is found in seafood and seaweeds. Other plants will contain iodine if grown on iodine rich soils. It is necessary for normal physical and mental growth and development, as well as for lactation and reproduction. An iodine deficiency is called goiter.

Iron - Occurs in plant and animal tissue. The body retains iron very well and only trace amounts are needed in diet. Iron is essential to form the oxygen-carrying hemoglobin in red blood cells and involved in muscle function.

Manganese - Plants are the best source for manganese. Trace amounts are necessary for healthy bones and for enzyme reactions involved in energy production.

Zinc - Zinc is found primarily in animals, but also occurs in plants growing on good soil. It is important for various enzyme reactions, the reproductive system, and for the manufacture of body protein.

Categories of Foods

There are nine categories of plant foods discussed in this book. They include root vegetables, green vegetables, fleshy fruits, seeds, nuts, and grains, flowers, and inner bark. Root vegetables (i.e., tubers, corms, bulbs, rhizomes, true roots) include plants such as wild onions, camas, spring beauty, glacier lily, bitterroot, balsamroot, and knotweed. Roots are the storage organs high in carbohydrates. The greatest amount of energy from roots is available at the end of the growing season. These carbohydrates come in a variety of forms and flavors, and are not always readily digestible by humans. One type of carbohydrate found in some roots is inulin, which becomes sweet after cooking due to its conversion to fructose. If the skin of a plant's root is consumed, it can provide minerals and small amounts of vitamins.

Green vegetables include leaves, stems, shoots, and buds. Examples are fireweed (shoot and stem), lambs quarters, nettles, and mustard leaves. Many green vegetables are most palatable and digestible when they are young. Green vegetables are high in moisture, and often contain carotene, vitamin C, folic acid, and various minerals (e.g., iron, calcium, magnesium).

Fleshy fruits include serviceberry, gooseberries, currants, huckleberries, wild plums, cherries, and rose hips. Fleshy fruits are a good source of ascorbic acid, and contain high amounts of other nutrients such as calcium, Vitamin A, and folic acid.

Seeds, nuts, and grains are a good source of protein, fat, carbohydrates, vitamins, and minerals. Oils can also be rendered from these foods. Nuts are especially good sources of B vitamins, amino acids, and iron.

The cambium or inner bark of coniferous and deciduous trees and shrubs is another category of plant foods. The inner bark may be scraped off trees in the spring. Many species have high sap content. For example, maple sap is high in carbohydrate/sugar energy value for an inner bark food.

The final category of plant foods are the flowers. Rose petals, fireweed flowers, and mariposa lily buds are high in moisture. Flowers are low in proteins and fats, but some are rich in vitamin A (carotene) or vitamin C. There is little published information on the mineral content of flowers.

The nutritional value of plants changes with the seasons. During spring and summer, many plants are tender and rich in vitamins. Roots and tubers are high in carbohydrates and other nutrients. But as summer progresses, roots become less desirable because the stored energy is shifted to the aboveground parts. Fall is a time of nuts and berries, which provide a good source of protein. Roots again begin to store carbohydrates. Winter, however, can be bleak. The aboveground edibles may be limited to berries that have persisted into winter, bark and pine needles for teas, and inner bark. Teas can be restorative and do provide some food value. Teas can be upgraded into stews by adding insect larvae, birds, or mammals to make them more nutritious and sustaining.

The nutritional value of plants also depends on preparation methods. For example, cooking greens in two changes of water makes them more palatable but can reduce the nutritional value. Generally, the preferred order of preparation for plants foods is: raw, quick-cook or steamed, baked, then boiled. Frying is the least desirable cooking method since it destroys many useful vitamins and minerals.

PROPER IDENTIFICATION

Do not eat any plant if uncertain of its identity! Although fatal mistakes are not very common (generally a person becomes really sick), they are possible with a handful of plants found in the Sawtooth Country.

Know the poisonous members of the Carrot family (Apiaceae), especially poison hemlock (*Conium*) and water hemlock (*Cicuta*) which may be common in some areas and the toxic nightshades like tobacco (*Nicotiana*) and Jimson weed (*Datura*). Many other families contain poisonous plants and should take the time to learn about them. Bottom line here: ***THERE SHOULD BE NO DOUBT OF A PLANT'S IDENTITY BEFORE COMSUMPTION.***

If you see the following symbol attached to a plant, exercise caution.

GUIDELINES FOR GATHERING

Since there are no general rules for distinguishing an edible plant from an unsavory or poisonous species, one must identify a plant correctly before attempting to use it. Some books suggest that if you don't know a plant, you can eat a small quantity and wait to see if it has any adverse effects. This is a potentially serious mistake. For instance, if the unknown plant happens to be death camas (*Zigadenus*), not only would it cause much discomfort (such as a burning sensation in the mouth), it could kill you. Anyone who plans to search out and consume edible plants should exercise extreme caution. Correct identification of plants is necessary to avoid similar species or parts that may be unpalatable or poisonous. One of the best ways to learn about plants is to consult a knowledgeable botanist or qualified individual.

It is important to harvest plants with wisdom and respect. The uncontrolled harvesting of plants could severely damage delicate plant communities. In addition, it may be illegal to injure or uproot a living plant in some areas covered by this handbook (e.g., national parks and monuments). If a plant is rare or endangered, look for other edibles. If you are not in a survival situation, you should be even more frugal and considerate.

Also, be mindful of your own safety when gathering edible and useful plants. In Idaho and elsewhere, city, county, state, and federal agencies and organizations often spray chemicals to control noxious weeds, especially in areas where logging, mining, and grazing activities occur, or in developed campgrounds. While such chemicals may be considered to be "safe" - there are no guarantees. You should avoid collecting in areas affected by pollutants such as along roads or in drainages affected by mining activities.

THERE IS NO SUCH THING AS A SAFE TEST FOR EDIBILITY!

FOLLOWING SUCH INSTRUCTIONS ~~CAN~~ WILL KILL YOU!

KEY TO COMMON PLANT FAMILIES FOUND IN IDAHO'S SAWTOOTH COUNTRY

(please note, not all families are included here; a more technical flora will need to be consulted for some groups of plants) - check out *A FIELD GUIDE TO IDAHO MOUNTAIN PLANTS* by Ray S. Vizgirdas

1. Plants reproducing by spores; ferns and horsetails ----- **2**
1. Plants reproducing by seeds; typical flowering plants and gymnosperms ----- **3**

2. Stems hollow and jointed; leaves in whorls ----- **Horsetail Family (Equisetaceae)**
2. Stem solid; leaves fern-like ----- **Ferns**

3. Ovules and seeds borne on face of a scale, not enclosed in an ovary (e.g., fruits); evergreen tree ("pines") ----- **Gymnosperms in Pinaceae**
3. ovules and seeds in a fruit; typical flowering plants ----- **4**

4. Leaves usually parallel-veined; flower parts in threes or sixes ----- **5 ("Moncots")**
4. Leaves usually netted veined; flower parts in fours or fives; rarely in twos -----   **10 ("Dicots")**

5. Plants with petal-like perianth ----- **6**
5. Plants with perianth of chaffy scales or hairy bristles; rushes, grasses, sedges, and cattails ----- **8**

6. Ovary inferior ----- **7**
6. Ovary superior ----- **Lily Family (Liliaceae)**

7. Flower parts regular, alike in size and shape ----- **Iris Family (Iridaceae)**
7. Flower parts irregular, not alike ----- **Orchid Family (Orchidaceae)**

8. Plants growing in marshy places ----- **9**
8. Plants not confined to wet, marshy places ----- **Grass Family (Poaceae)**

9. Stems round or nearly so ----- **Cattail Family (Typhaceae)**
9. Stems appearing triangular because of the leaves in three rows ----- **Sedge Family (Cyperaceae)**

10. Petals absent ----- **11**
10. Petals present ----- **30**

11. Trees or shrubs ----- **12**
11. Herbaceous plants (maybe woody/shrubby at the base) ----- **16**

12. Flowers not in catkins ----- **13**
12. Flowers in catkins ----- **15**

13. Leaves opposite ----- **Maple Family (Aceraceae)**
13. Leaves alternate ----- **14**

14. Ovary with one cell ----- **Elm Family (Ulmaceae)**
14. Ovary with two or four cells ----- **Buckthorn Family (Rhamnaceae)**

15. Calyx present; ovary with one or two ovules or seeds; leaves without stipules ----- **Birch Family (Betulaceae)**
15. Calyx absent; ovary with many ovules or seeds; leaves usually with stipules ----- **Willow Family (Salicaceae)**

16. Leaves opposite ----- **17**
16. Leaves not opposite ----- **20**

17. Flowers perfect (having stamens and pistiles); plants without stinging hairs ----- **18**
17. Flowers not perfect; plant with stinging hairs ----- **Nettle Family (Urticaceae)**

18. Sepals united into a corolla-like tube ----- **Four-o'clock Family (Nyctaginaceae)**
18. Sepals not united, but free to base ----- **19**

19. Fruit a many-seeded capsule; leaves opposite ----- **Pink Family (Caryophyllaceae)**
19. Fruit an achene; leaves seldom opposite ----- **Buckwheat Family (Polygonaceae)**

20. Flowers with stamens and pistils in separate flowers on the same plant; surface of plant usually scurfy or mealy ----- **Goosefoot Family (Chenopodiaceae)**
20. Flowers perfect, or if not perfect, then staminate and pistillate flowers are on separate plants; leaves not mealy ----- **21**

21. Pistils more than one ----- **22**
21. Pistils only one ----- **23**

22. Receptacle cup-shaped with stamens attached on the rim of the cup and above the ovary ----- **Rose Family (Rosaceae)**
22. Receptacle cone-like or flat, but never cup-shaped; stamens attached below or at base of the ovary ----- **Buttercup Family (Ranunculaceae)**

23. Ovary more than one-celled ----- **Mustard Family (Brassicaceae)**
23. Ovary only one-celled ----- **24**

24. Ovary wholly superior ----- **25**
24. Ovary partly inferior ----- **28**

25. Sepals and petals four each, stamens 6 ----- **Caper Family (Capparaceae)**
25. Sepals and petals not four each, stamens not six ----- **26**

26. Leaves with papery sheathing stipules, especially in young stems (except Eriogonum, no stipules) ----- **Buckwheat Family (Polygonaceae)**
26. Leaves without sheathing stipules ----- **27**

27. Fruit an achene; leaves not mealy ----- **Buckwheat Family (Polygonanceae)**
27. Fruit not an achene, but a utricle; leaves often mealy ----- **Goosefoot Family (Chenopodiaceae)**

28. Flowers on a leafy stem ----- **Sandalwood family (Santalaceae)**
28. Flowers on stem with few or no leaves ----- **Saxifrage Family (Saxifragaceae)**

30. Petals separate, not united to each other to form a tubular corolla ----- **31**
30. Petals united to each other at least at the base, to form a tubular corolla ----- **61**

31. Ovary superior ----- **32**
31. Ovary wholly or partly inferior ----- **56**

32. Stamens same number as the petals and opposite them ----- **33**
32. Stamens either not the same number as the petals, or alternate with them ----- **36**

33. Shrubs or small trees ----- **34**
33. Herbaceous plants ----- **35**

34. Leaves compound ----- **Barberry Family (Berberidaceae)**
34. Leaves simple ----- **Buckthorn Family (Rhamnaceae)**

35. Sepals 2; plants fleshy or succulent ----- **Purslane Family (Portulacaceae)**
35. Sepals 6; plants not succulent ----- **Barberry Family (Berberidaceae)**

36. Pistils one to many, but always simple, as indicated by only one stigma, style, or placenta ----- **37**
36. Pistil 1, but always compound, as shown by the two or more stigmas, styles, locules or placentae ----- **43**

37. Pistil one ----- **38**
37. Pistil 2 to many ----- **39**

38. Flowers irregular, papilionaceous ----- **Pea Family (Fabaceae)**
38. Flowers regular ----- **Rose Family (Rosaceae)**

39. Sepals not united to each other ----- **40**
39. Sepals united to each other to form a floral cup around the ovary ----- **41**

40. Stamens 10 ----- **Stonecrop Family (Crassulaceae)**
40. Stamens many (>10) ----- **Buttercup Family (Ranunculaceae)**

41. Stamens 5-12, usually 10 ----- **Saxifrage Family (Saxifragaceae)**
41. Stamens many ----- **42**

42. Stamens united into one or more series around the pistils ----- Mallow Family (Malvaceae)
42. Stamens free and distinct, not united ----- Rosaceae

43. Ovary deeply 5-lobed with one seed in each lobe; or ovaries united in a ring around a central axis ----- **44**
43. Ovary not appearing to be 5-lobed, nor united in a ring around the central axis ----- **45**

44. Stamen filaments united into a tube; carpels 5-9; ovaries united in a ring around a central axis ----- **Mallow Family (Malvaceae)**
44. Stamen filaments not united to each other ----- **Geranium Family (Geraniaceae)**

45. Sepals 2 ----- **Purslane Family (Portulacaceae)**
45. Sepals 4 to 5 ----- **46**

46. Sepals not united to each other ----- **47**
46. Sepals united to each other ----- **52**

47. Sepals 5; petals 5 ----- **48**
47. Sepals four; petals four ----- **Mustard Family (Brassicaceae)**

48. Flowers irregular ----- **Violet Family (Violaceae)**
48. Flowers regular ----- **49**

49. Stamens united with the base of the corolla; stamens more numerous than corolla lobes; ovary 3 to many celled ----- **Mallow Family (Malvaceae)**
49. Plants not as above ----- **50**

50. Flowers pink or red ----- **Geranium Family (Geraniaceae)**
50. Flowers white or blue ----- **51**

51. Flowers white ----- **Pink Family (Caryophyllaceae)**
51. Flowers blue -----**Flax Family (Linaceae)**

52. Leaves alternate or basal ----- **53**
52. Leaves opposite ----- **55**

53. Woody vines or shrubs ----- **Sumac Family (Anacardiaceae)**
53. Herbs ----- **54**

54. Stamens 10 or less ----- **Saxifrage Family (Saxifragaceae)**
54. Stamens numerous (>10) ----- **Rose Family (Rosaceae)**

55. Fruit a 2-winged samara ----- **Maple Family (Aceraceae)**
55. Fruit a capsule ----- **Pink Family (Caryophyllaceae)**

56. Flowers in umbels ----- **Carrot Family (Apiaceae)**
56. Flowers not in umbels ----- **57**

57. Stamens the same number as the petals and opposite them ----- **Saxifrage Family (Saxifragaceae)**
57. Stamens not the same number as the petals and alternate with them ----- **58**

58. Calyx lobes and petals five each ----- **59**
58. Calyx lobes and petals four each ----- **60**

59. Stamens many ----- **Rose Family (Rosaceae)**
59. Stamens 5-10 ----- **Saxifrage Family (Saxifragaceae)**

60. Shrubs ----- **Dogwood Family (Cornaceae)**
60. Herbs ----- **Evening-primrose Family (Onagraceae)**

61. Stamens more than 5 ----- **62**
61. Stamens five or less ----- **63**

62. Stamens many, their filaments united into a tube ----- **Mallow Family (Malvaceae)**
62. Stamens 6, united into two sets of 3 each ----- **Fumatory Family (Fumariaceae)** Dicentra needs to be included

63. Ovary superior ----- **64**
63. Ovary inferior ----- **76**

64. Corolla irregular (*Veronica* in Scrophulariaceae is slightly irregular) ----- **65**
64. Corolla not at all irregular ----- **68**

65. Leaves alternate or basal, or only the lower ones opposite ----- **66**
65. Leaves all opposite or whorled ----- **67**

66. Root parasites without chlorophyll or normal foliage ----- **Broom-rape Family (Orobanchaceae)**
66. Green plants with normal foliage, not parasites ----- **Figwort Family (Scrophulariaceae)**

67. Mature ovary deeply lobed or divided into four 1-seeded nutlets; style 2-branched; stems square; one seed per cell of the ovary ----- **Mint Family (Lamiaceae)**
67. Mature ovary a capsule, not lobed into nutlets, but may be notched so that it appears somewhat lobed; stems round; several seeds per cell of ovary ----- **Figwort Family (Scrophulariaceae)**

68. Ovary divided or lobed into 4 parts around the base of the style ----- **Borage Family (Boraginaceae)**
68. Ovary not 4-lobed around the style ----- **69**

69. Stamens opposite the corolla lobes ----- **Primrose Family (Primulaceae)**
69. Stamens attached between the corolla lobes (alternate) ----- **70**

70. Twinning or trailing herbs ----- **71**
70. Plants not twinning or trailing ----- **72**

71. Sepals separate to the base ----- **Morning Glory Family (Convolvulaceae)**
71. Sepals united ----- **Nightshade Family (Solanaceae)**

72. Ovary 1-celled ----- **73**
72. Ovary 2 or more celled ----- **74**

73. Leaves entire, opposite or whorled ----- **Gentian Family (Gentianaceae)**
73. Leaves, if entire, alternate or basal ----- **Waterleaf Family (Hydrophyllaceae)**

74. Stamens extending beyond the corolla ----- **Waterleaf Family (Hydrophyllaceae)**
74. Stamens shorter than the corolla ----- **75**

75. Style unbranched ----- **Nightshade Family (Solanaceae)**
75. Style branched ----- **Phlox Fmily (Polemoniaceae)**

76. Flowers in heads surrounded by several leaf-like bracts; anthers united in a ring or tube around the pistil ----- **Sunflower Family (Asteraceae)**
76. Flowers not as above ----- **77**

77. Shrubs or woody vines ----- **Honeysuckle Family (Caprifoliaceae)**
77. Herbs ----- **78**

78. Stamens 1-3; upper leaves apposite ----- **Valerian Family (Valerianaceae)**
78. Stamens 4 or 5; leaves whorled ----- **Madder Family (Rubiaceae)**

HANDBOOK TO IDAHO'S SAWTOOTH COUNTRY FLORA

MAPLE
Acer   ACERACEAE

General Description Maples are deciduous trees or shrubs with male and female flowers on the same or separate plants. Flowers are arranged in racemes, corymbs, or umbels. Fruits are winged schizocarps that resemble tiny "helicopters" when blown by the wind. Within the Sawtooth Country, two species may be encountered and include Rocky Mountain maple (*A. glabrum*) and boxelder (*A. negundo*).

1. Leaves compound with 3-5 leaflets, terminal leaflet stalked ----- ***A. negundo***
1. Leaves simple ----- ***A. glabrum***

Ecology & Ethnobotany Like with other maples, syrup can be extracted by boiling down the sap. The winged seeds can be roasted for food. The young shoots and inner bark are valuable in times of emergency as food – dried and ground into flour.

Maple wood has been used to make snowshoe framing, mush paddles, and other household utensils. Knots

and burls on tree trunks can be used for making bowls, dishes, and other items. Inner bark can be shredded and twisted into a coarse rope.

WATER-PLANTAIN
Alisma triviale   ALISMATACEAE

General Description These are perennial plants arising from fleshy, bulb-like stems. The basal leaves are long stalked and egg-shaped, and the flowers are white. Water-plantain is usually found in marshes and ponds at lower elevations.

Ecology & Ethnobotany The starchy, bulbous bases of water-plantain are edible as a starchy vegetable (potato) after drying. Drying is said to remove the strong flavor. *Alisma* has a long history of use in Chinese medicine and is mentioned in texts dating back to about 200 A.D. It was also used by early herbalists as a diuretic and by the Cherokee Indians for application to sores, wounds, and bruises. It is described as a sweet, cooling herb that lowers blood pressure, cholesterol and blood sugar levels.

The necessities of life are all around you, but they are not there for your benefit. They are not packaged and labeled for your convenience, and no sheet of instructions comes with them. You will have to find and procure what you need by your own unaided effort, and you must adapt it to your use on the basis of your own knowledge.

AMARANTH, PIGWEED
Amaranthus   AMARANTHACEAE

General Description In general, these are herbaceous annuals with small greenish flowers, and alternate entire or wavy margined leaves. Within the Sawtooth's at least 2 species of *Amaranthus* may be encountered. They include: prostrate pigweed (*A. albus*) and mat amaranth (*A. blitoides*), The following key may be helpful in identifying the species.

1. Sepals 3 ----- ***A. albus***
1. Sepals 4 or 5 ----- ***A. blitoides***

Ecology & Ethnobotany The seeds of all species can be eaten whole as a cereal or ground into meal and made into cakes. The seeds are best collected in summer when the plants are fully mature. To free the seeds from their husks, rub the seed clusters between your hands. You can then winnow the seeds if there is a breeze, or if air is calm, slowly pour the seeds out of your hands and blow the chaff away. The seeds contain approximately 15 grams of protein per 100 grams, more than is found in rice and corn, and equal to, if not surpassing that found in wheat. When ground into a flour, amaranth has a distinctive flavor that is a bit strong used alone. We find it is better when mixed with other flours for breads and pancakes.

The highly nutritious amaranth contains more fiber and calcium than any other cereal grain in addition to a wide spectrum of vitamins and minerals, including Vitamins A and C, calcium, magnesium, and iron. Amaranths is rich in the amino acid lysine, a product scarce in true cereal grains, thereby providing a more balanced source of protein.

The edible young shoots and leaves have a pleasant taste if eaten as a potherb soon after collection. Since the plants can accumulate nitrates, it is wise not to consume large quantities where nitrate fertilizers are used.

FRAGRANT SUMAC
Rhus aromatica   ANACARDIACEAE

General Description Also known as skunkbush, this is a shrub that grows to 5 feet tall has 3-foliate leaves and yellowish flowers. Blooms from March to April. Skunkbush is one of the plants that shows measle-like splotching in the presence of ozone. As such, it is being watched by scientists in some areas as an indicator of air pollution. The plant previously classified as Rhus trilobabta.

Ecology & Ethnobotany The berries can be steeped in water to make an interesting tasting drink. The slender, flexible branches of skunkbrush can be used for weaving baskets as they are somewhat vine-like. The branches can also be used as chew-sticks to clean teeth and massage gums. Take a small stem several inches long, remove the outer bark and chew on the tip to soften fibers. *Since some people may have an allergic reaction to the oils of sumac, it is recommended that this be done sparingly.*

WATER HEMLOCK
Cicuta   APIACEAE

General Description Water hemlock is a glabrous perennial with stout stems up to 5 feet tall that have hollow, tuberous-thickened bases with horizontal cross sections inside. Leaves are 1-2 times pinnately divided into narrowly lance-shaped, sharply toothed, and evidently veined leaflets 1-4 inches long. Greenish to white flowers are borne in compound umbels with rays $\frac{1}{4}$ to $\frac{3}{4}$ inch long. The fruits, less than 1/8- inch long, are slightly flattened with thickened ribs on the faces. The bruised foliage produces a musky odor. The plant can be found in marshes and along the edges of streams and ponds from low to mid-elevations.

Ecology & Ethnobotany These are *extremely poisonous plants if ingested*. In fact, water hemlock has been described as the *most violently poisonous vascular plant in North America.* The whole plant contains cicutoxin, a resin-like substance that depresses the respiratory system, with the root being particularly dangerous. A single mouthful of the plant is capable of killing an adult. Water hemlocks have been used throughout the ages to execute criminals and kings. Many children have been fatally poisoned by blowing into whistles made from hollow stems of water hemlock. It is imperative that you take extra special care in your ability to identify members of the carrot family (Apiaceae). The following should reinforce this fact:

"When about the end of March 1670, the cattle were being led from the village to water at the spring, in treading the river banks they exposed the roots of this Cicuta (water hemlock) whose stems and leaf buds were now coming forth. At that time two boys and six girls, a little before noon, ran out to the spring and the meadow through which the river flows, and seeing a root and thinking that was a golden parsnip, not through the bidding of any evil appetite, but at the behest of wayward frolicsomeness, ate greedily of it, and certain of the girls among them commended the root to others for its sweetness and pleasantness, wherefore the boys, especially, ate quite abundantly of it and joyfully hastened home; and one of the girls tearfully complained to her mother she had been supplied meagerly by her comrades, with the root.

Jacob Maeder, a boy of six years, possessed of white locks, and delicate though active, returned home happy and smiling, as if things had gone well. A little while afterwards he complained of pain in his abdomen, and, scarcely uttering a word, fell prostrate to the ground, and urinated with great violence to the height of a man. Presently he was a terrible sight to see, being seized with convulsions, with the loss of all his senses. His mouth was shut most tightly so that it could not be opened by any means. He grated his teeth; he twisted his eyes about strangely and blood flowed from his ears. In the region of his abdomen a certain swollen body of the size of a man's fist struck the hand of the afflicted father with the greatest force, particularly in the neighborhood of the ensiform cartilage. He frequently hiccupped; at times he seemed to be about to vomit, but he could force nothing from his mouth, which was most tightly closed. He tossed his limbs about marvelously and twisted them; frequently his head was drawn backward and his whole back was curved in the form of a bow, so that a small child could have crept beneath him in the space between his back and the bed without touching him. When the convulsions ceased momentarily, he implored the assistance of his mother. Presently, when they returned with equal violence, he could not be aroused by no pinching, by no talking, or by no other means, until his strength failed and he grew pale; and when a hand was placed on his breast he breathed his last. These symptoms continued scarcely beyond a half hour. After his death, his abdomen and face swelled without lividness except that a little was noticeable about the eyes. From the mouth of the corpse even to the hour of his burial green froth flowed very abundantly, and although it was wiped away frequently by his grieving father, nevertheless new froth soon took its place."

☠ POISON HEMLOCK
Conium maculatum   APIACEAE

General Description Poison hemlock is a biennial with a stout taproot and a disagreeable odor when crushed. The stems are purple-blotched and hollow, and the leaves are pinnately dissected with a lacy appearance to them. The flowers are white in compound umbels, and the fruits are egg-shaped, flattened with prominent, wavy ribs. The plant is usually found in disturbed sites and waste places at low elevations (below 5,000 feet). Blooms April to September.

Ecology & Ethnobotany This is an *extremely poisonous plant*. Death is said to result from the ingestion of the leaves, roots or seeds. The most famous use of poison hemlock was by the ancient Greeks as a humane method of capital punishment. It is said to be quite painless and the recipient's mind remains clear to the end. Introduced from Europe, poison hemlock has established itself as a common weed.

SPRINGPARSLEY
Cymopterus　　**APIACEAE**

General Description These are low perennial herbs with long, thick taproots. The leaves are 2-4 times pinnately divided into small ultimate segments, and the flowers are yellow or white in terminal, compound umbels. The round fruits have winged ribs on the outer faces. Most of the species occur in dry soils or gravelly slopes at the lower elevations.

Ecology & Ethnobotany All species produce edible roots. We found the older roots more fibrous than the younger ones. The root can be used in stews or it can be boiled or roasted in a pit, mashed and dried into cakes. When dried, it will keep indefinitely. During the Lewis and Clark expedition this was known as *kouse* (bread of cows). The old roots can also be used as an effective insect repellant when boiled. Just sprinkle the tea around camp and in sleeping areas.

The upper parts of the plants have been used raw or as potherbs. If cooked, they will require several changes of water.

COMMON COWPARSNIP
Heracleum maximum   APIACEAE

General Description This is a stout perennial growing up 7 feet tall. The lower leaves are three lobed, resembling a maple leaf up to 14 inches long. The white flowers occur in compound umbels and the fruits are egg-shaped with only the marginal ribs winged. In the Sawtooth Country, cowparsnip is usually found in moist soils around streams, seeps, and avalanche chutes up to subalpine environments. It blooms April to June.

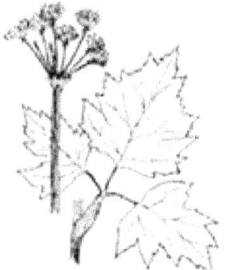

Ecology & Ethnobotany The young stems of cowparsnip can be peeled and eaten raw but are best when cooked. The hollow base of the plant can be cut into short lengths and used as a substitute for salt by eating or cooking with other foods. The young leaves are also edible after cooking, but we find them not very tasty. The leaves can also be dried and burned, and the ashes used as a salt substitute. Strong and bitter tasting, the cooked roots are said to be good for digestion, as well as for relieving gas, and cramps. In our experience, some plants are much more palatable than others.

The seeds can be sparingly added to salads for seasoning. However, you should be aware that the mature, green seeds have a mild anesthetic action on tissues in the mouth. When gently chewed and sucked, they will numb the tongue and gums in a manner similar to clove oil.

The leaves of cowparsnip are large enough to be used as a toilet paper substitute and as a mild insect repellent. However, since furanocoumarin is present in the sap and the outer hairs, it may be a problem for people with sensitive skin. When the sap comes in contact with the skin in sensitive people it causes a type of "sunburn" effect (i.e., redness, blistering, and running sores) when exposed to light.

The older stems, before the flower cluster unfolds, can be peeled and the inner tissue eaten raw or cooked. While it is edible, it does have an unpleasant taste. Cooking it in a couple changes of water usually helps the taste and digestibility. In any case, cowparsnip is considered to be an excellent survival plant in the mountains. **Warning** Do not confuse this plant with other species in the same family that are highly toxic (i.e., *Cicuta* and *Conium maculatum*)

LICORICE-ROOT
Ligusticum   **APIACEAE**

General Description At least 3 species may be encountered within the Sawtooth Country: *L. filicinum* (fernleaf licorice-root), *L. grayi* (Gray's licorice-root), and *L. tenuifolium* (Idaho licorice-root).

They are perennials with taproots sheathed by old leaf bases at the crown. The leaves are 1-3 times pinnately dissected, the white to pinkish flowers are arranged in compound umbels. Fruits are oblong to elliptical with winged ribs. Look for these plants in moist habitats.

Ecology & Ethnobotany Native Americans in California cooked *Ligusticum* leaves in water and then in earth ovens before eating. The tender leaves of *L. grayi* were soaked in water, cooked, and then used as a meat substitute when acorns were eaten.

Medicinally, some Native Americans used crushed *Ligusticum* roots to poultice sprains and bruises. Additionally, pulverized roots were sprinkled into queit pools to stun fish, making them easier to catch.

WILD PARSLEY, BISCUITROOT
Lomatium   APIACEAE

General Description These are perennial plants with thick roots and leaves that are divided several times from the base. The white, yellow, pink, or purplish flowers are in compound umbels. The fruits are flattened and elliptical to oval in shape, and the margins may or may not be winged. At least 13 species of biscuitroot may be encountered within the Sawtooth Country. Look for them in dry ground or rocky situations.

Ecology & Ethnobotany All species have edible roots and were an important staple among many Native Americans. They can be eaten raw

or cooked, or dried and ground into flour. The flour can then be kneaded into dough, flattened into cakes, and dried in the sun or baked. Some of the species we have tried were too resinous to enjoy. Personal taste will guide one to choose the more palatable species.

The green stems can be eaten after boiling in the springtime, but as summer progresses they become tough and fibrous. A tea can be brewed from leaves, stems, and flowers. The tiny seeds nutritious raw or roasted, can be dried and ground into meal. The plants are rich in Vitamin C.

GREAT BASIN INDIAN POTATO
Orogenia linearifolia   APIACEAE

General Description This is a small perennial with fleshy roots. Flowers are white and in compound umbels. Look for the species soon after snows melt in the mountains in spring and early summer. They are sometimes called Snow Drops. The species may be elusive as it is normally found early in the spring.

Ecology & Ethnobotany The roots can be boiled, steamed, roasted or baked in any way used for preparing potatoes. When small, they can be eaten raw. They can also be cooked and mashed into cakes for drying, and when protected from moisture, will keep a long time. The hard cakes can be soaked and cooked in stews.

SWEETCICELY, SWEETROOT
Osmorhiza   APIACEAE

General Description These are herbaceous perennials from stout roots, with leaves twice divided into 3's. The flowers are borne in open, compound umbels that arise from leaf axils. The fruit is spindle-shaped and compressed along the sides. At least four species may be encountered in the Sawtooth Country on moist slopes, open areas, and in forests. They include: sweetcicely (*O. berteroi*), bluntseed sweetroot (*O. depauperata*), western sweetroot (*O. occidentalis*), and purple sweetroot (*O. purpurea*). The following key may be useful in distinguishing the species.

1. Flowers yellow and fruits glabrous; plants smelling of anise ----- ***O. occidentalis***
1. Flowers white to pink or purple; fruits hairy; plants not strongly odoriferous ----- **2**

2. Fruits hairy below but glabrous on the upper third ------ ***O. purpurea***
2. Fruits hairy to near the tip ----- **3**

3. Fruits concavely narrowed to a beak-like tip; flowers white ------ ***O. berteroi***
3. Fruits abruptly rounded at the tip, beak lacking; flowers sometimes purplish or pink ----- ***O. depauperata***

Ecology & Ethnobotany The roots and fruits of any species should be tried for food. The roots in particular are recorded as being heavy with a sweet licorice or anise-like flavor, and where this is too strong, the roots and seeds can be dried and pulverized for use as a food flavoring.

YAMPAH
Perideridia   **APIACEAE**

General Description At least two species of yampah can be encountered within the Sawtooth Country and include Bolander's yampah (*P. bolanderi*) and Gardner's yampah (*P. gairdneri*). They are biennial or perennial herbs with fascicled tuberous roots and pinnate leaves. The calyx-teeth are well-developed. The petals are white or pinkish, the stylopodium conic. The fruit is nearly terete or somewhat flattened laterally. The following key may be useful in distinguishing the species.

1. Main leaves are somewhat dissected; petioles dilated; fruit oblong in shape ----- ***P. bolanderi***
1. Main leaves are only once or maybe twice pinnate or ternate; petioles not dilated; fruit oval in shape ----- ***P. gardneri***

Ecology & Ethnobotany All of the species within this genus are edible. They were an important food of many indigenous peoples from British Columbia to California and the Great Basin region. The raw finger-like roots have a pleasant, nutty flavor when eaten raw, and resemble carrots when cooked. They are best when dug up before the flowers appear. The roots should be washed

and peeled before cooking. They can be easily dried and will keep well for future use. When dried, the roots can be pounded and ground into flour or mashed into cakes. The seeds may be parched and ground or eaten whole.

 HEMLOCK WATER PARSNIP
Sium suave   APIACEAE

General Description This is a stout plant up to five feet tall. The leaves are pinnately divided, lower leaves with long petioles while upper leaves almost sessile. Veins of leaflets not in line with teeth or notches below teeth. Flowers are white, and the fruit is oval in shape. Within the Sawtooth Country the plants is usually found in water or swampy areas in the mountains.

Ecology & Ethnobotany Edible, but best avoided due to its similarity to the deadly piosonous water wemlock (*Cicuta maculata*) - see warning below.

The long fleshy root of water parsnip, which is edible raw or cooked, has a sweet, carrot-like taste. The leaves and younger stems are also edible after cooking, but we found them better when boiled until tender. The older plants and flowers should be avoided because they are toxic and have been suspected of poisoning a wide range of livestock.

Warning *The plant is very similar in form and habitat to* Cicuta maculata *(water hemlock) which is the most poisonous vascular plant in North America.* Both plants produce white flowers in umbrella-like clusters and both grow in swampy ground at lake or pond edges. Water parsnip has leaves that are once-compound, whereas the leaves of water hemlock are 3-times compound. Water hemlock also has a distinctive turnip-like swelling at the base of the stem, which is usually chambered when cut open vertically and exudes a yellowish liquid along the cut surface. *Therefore, when in doubt, leave it alone!!!*

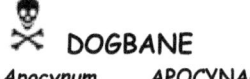
DOGBANE
Apocynum   APOCYNACEAE

General Description Two species may be encountered within the Sawtooth Country: spreading dogbane (*A. androsaemifolium*) and Indian hemp (*A. cannabinum*). These are perennial herbs with milky juice, that have leaves that are opposite, and the pink, bell-shaped flowers are borne in cymes. The following key may be useful in distinguishing the two species.

1. Flowers pinkish in color ----- ***A. androsaemifolium***
1. Flowers greenish in color ----- ***A. cannabinum***

Ecology & Ethnobotany Dogbanes should be considered poisonous to humans if ingested. The primary use of dogbanes is for fiber. The stem fibers are strong and can be used for rope making, mats, baskets, bowstrings, fishing lines and nets, sewing, animal trap triggers, snares, cordage for bow and drill fire making, and general weaving. One of the easiest ways to isolate the fibers is to soak the stems in water.

# MILKWEED
Asclepias   ASCLEPIADACEAE

General Description Three species of milkweed could be encountered within the Swtooth Country. They include: pallid milkweed (*A. cryptoceras*), Mexican whorled milkweed (*A. fascicularis*), and showy milkweed (*A. speciosa*). In general, they are erect or decumbent herbs from deep perennial roots. The leaves are opposite or whorled and the corolla is deeply 5-parted with the segments reflexed. The corona hoods each have an incurved horn within. The larvae of the monarch butterfly (*Danaus plexippus*) feeds on the leaves of milkweeds. The following key may be useful in distinguishing the species.

1. Leaves linear to linear-oblong in shape, about ¼ inch or less wide ----- *A. fascicularis*
1. Leaves lanceolate or broader in shape ----- **2**

2. Leaves oval in shape, barely longer than wide; plants more or less glabrous; flowers large, greenish yellow or white in color; follicles (fruits) ovate in shape ----- *A. cryptoceras*
2. Leaves usually twice or more as long as wide; plants hairy ----- *A. speciosa*

Ecology & Ethnobotany These are another group of important fiber plants for hikers. The strong fiber can be obtained from the inner bark to make rope, fishing line, clothing, and nets. Archeologists have discovered clothing that was made from the fibers more than 10,000 years ago. The silky floss found in mature milkweed seed pods were used in making candlewicks, and the fiber can be spun like cotton. The floss is buoyant and water resistant and makes a good insulator. The dried pods were used as utensils. The sap was used as an adhesive.

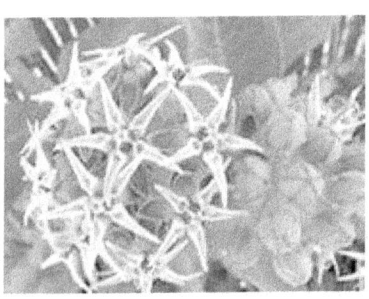

Down from cattails, fireweed, and milkweed, and thistles is very high in insulation value, but tends to lose this quality rather quickly as it mats or packs. Use such plant-derived down with harsher and more mat-resistant materials for best results. Dead grasses, dry leaves, moss, and dead pine needles are good for this when used alone if they are first thoroughly dried and then crumpled to make them occupy more space.

YARROW
Achillea millefolium   ASTERACEAE

General Description This is a strongly scented perennial herb with alternate leaves that are finely dissected and appear feathery. The white or sometimes yellow flowers are borne in a flat-topped corymb. Yarrow is widespread and can be found in a variety of habitats from low elevations to above timberline.

Ecology & Ethnobotany Yarrow is often referred to as "poor man's pepper." The leaves can be dried, ground, and used as seasoning. The young leaves can be added to salads. The aromatic leaves were also placed in freshly split fish to expedite drying. Rubbing the plant on one's clothing and skin, was an ancient prescription for repelling biting insects. The stalks burned on coals were said

to deter mosquitoes. The leaves were used in herbal snuffs and smoking tobaccos. Yarrow has also been used as a hops substitute for brewing yarrow beer.

MOUNTAIN DANDELION
Agoseris   ASTERACEAE

General Description These are annual or perennial, tap-rooted herbs with milky juice that resemble the common dandelion (*Taraxacum*). The flowers are all ray, yellow or occasionally orange in color. The pappus is white with hair-like bristles. The fruit (achene) is conspicuously ten-nerved. Mountain dandelions occur on moist to dry ground, in meadows and open areas at various elevations within the Sawtooth Country. They include orange agoseris (*A. aurantiaca*), pale agoseris (*A. glauca*), bigflower agoseris (*A. grandiflora*), annual agoseris (*A. heterophylla*), and spearleaf agoseris (*A. retrorsa*). The following key may be useful in identifying the species.

1. Beak 2-4 times as long as the body of the achene; flowers yellow, sometimes drying pink ----- **2**
1. Beak less than twice as long as the body of the achene ----- **3**

2. Lobes on leaves pointing strongly backward; achene abruptly beaked from the summit of achene ----- ***A. retrorsa***
2. Lobes on leaves not pointing strongly backward; achene tapering to the beak ----- ***A. grandiflora***

3. Flowers yellow, turning pink when dry ----- **4**
3. Flowers orange to pink or purplish ----- ***A. aurantica***

4. Plants annual, with weak taproot ----- ***A. heterophylla***
4. Plants perennial, with stout taproots ----- ***A. glauca***

Ecology & Ethnobotany The leaves and roots of some species are edible when cooked but are bitter, especially in late season. The seeds were eaten by the Native Americans. The sap from the leaves of some species, when hardened can be used as chewing gum. Since the sap from some species is very thick and insoluble, it may be useful for waterproofing containers (e.g., coiled baskets) and footwear.

WESTERN PEARLY-EVERLASTING
Anaphalis margaritacea   ASTERACEAE

General Description This is a rhizomatous perennial with distinctive white, woolly leaves and stems. The flowering heads are composed of disk flowers with yellow flowers surrounded by conspicuous, papery white involucral bracts. The pappus is comprised of capillary bristles. Pearly-everlasting can be found in various habitats from the foothills to above timberline.

Ecology & Ethnobotany The herbage of western pearly-everlasting has been used as a tobacco substitute to relieve headaches. As a tea, the plant has been used for colds, bronchial coughs, and throat infections. The whole plant can be used as a wash or poultice for external wounds. It has also been used for rheumatism, burns, sores, bruises, and swellings.

EVERLASTING, PUSSYTOES
Antennaria   ASTERACEAE

General Description These are herbaceous often mat-forming perennials. The heads are discoid, with small, white flowers surrounded by bracts that are typically hairy below with a smooth and membranous portion varying in color from white to pink to dark brown or black. The pappus is composed of numerous hairy bristles. At least 12 species may be encountered in dry, open habitats, or in moist or seasonally wet places from the foothills to alpine areas in the Sawtooth Country.

Ecology & Ethnobotany The sap from the stem of most species can be chewed like gum and has some nutritive value. Leaves can be used as a poultice for use on bruises, sprains, and swelling. The blossoms could be boiled and used to bathe sore or ulcerated feet, or mashed and applied to sores.

COMMON BURDOCK
Arctium minus   ASTERACEAE

General Description Common burdock is one of two species that may be encountered in the Sawooth Country. Introduced from Europe, it is a coarse biennial that grows up to five feet tall. The leaves are heart-shaped, with the lower surface slightly hairy, the upper surface glabrous. The heads are discoid with purple flowers. The narrow, hook-tipped involucral bracts spread when in fruit to form the familiar "sticky" burs. Common burdock is a familiar weed of waste places, usually found at the lower elevations. *Arctium lappa* (greater burdock) also occurs and can be used in much the same way as common burdock.

Ecology & Ethnobotany Rich in vitamins and iron, the young leaves and shoots can be gathered for use as a potherb or eaten raw in salad. The plant has a strong rank taste and an objectionable odor. The inner pith-like material of the young stems can be eaten raw, but we find it better when boiled in one or two changes of water. The roots of young plants can be sliced and cooked, then eaten. The older roots can be roasted and ground for use as a tea or coffee substitute. Seeds can be dampened and grown as sprouts.

The tall rigid stems can be used as drills for primitive fire-starting techniques. The burs can be used as a survival "velcro" for holding clothes together.

☠ ARNICA
Arnica   ASTERACEAE

General Description These are perennials arising from a rhizome or caudex. The leaves are simple and opposite. Flower heads are composed of ray and disk flowers, which are usually yellow or orange. The green involucral bracts occur in one series and the pappus consists of fine white or tawny bristles. At least 12 species of may be encountered in the Sawtooth Country. Arnicas are usually found in forested or open areas up to the subalpine zone.

Ecology & Ethnobotany All the species are reported to be poisonous if taken internally. arnica contains arnicin, choline, a volatile oil, arnidendiol, angelic and formic acid, and other unidentified substances that can alter cardiovascular activity. The Federal Drug Administration considers arnica as "unsafe" and bans its use for human consumption. The chopped plant is steeped in rubbing alcohol for about a week and squeezed through a cloth. The liniment is then used for joint inflammations, sprains, and sore muscles.

It should not be used if the skin is broken since it is toxic if it enters the bloodstream. Wear gloves as the volatile oils can be absorbed. **Warning** All species of arnica are reported to be poisonous if taken internally.

GREAT BASIN SAGEBRUSH
Artemisia tridentata   ASTERACEAE

General Description This is an evergreen shrub grows 1 to 9 feet tall and has silvery gray herbage. The leaves are narrow and spatulate, 3/8 to 1½ inches long, and with a 3-toothed apex. The inflorescence is 6 to 16 inches long, and composed of many, small, greenish flower heads. The involucres are 1/8-inch long and there are no ray flowers. There are 3 to 16 disk flowers per head, and there is no pappus. Basin sagebrush is quite common on dry slopes and plains. Blooms from August to October.

Ecology & Ethnobotany The seeds are edible raw or as flour. Since many species are aromatic, they can be used to store buried food caches by masking the odor of foodstuff, and to rub on the body to mask human scent while hunting. The wood of *A. tridentata* is a good material for fire drills. Although cordage can be made from the bark, it is not very strong.

MUGWORT, WORMWOOD
Artemisia   ASTERACEAE

General Description There are a number of species of *Artemisia*, including annual, biennial, and perennial herbs and shrubs. They are mostly aromatic with entire or dissected leaves. The flower heads are small, inconspicuous, and comprised of disk flowers. The following three species are commonly encountered in the Sawtooth Country.

Mugwort (*A. douglasiana*) is a perennial growing 20 to 60 inches tall, and has leaves, that are 2 to 6 inches long, lanceolate to elliptic in shape, and entire or few-toothed; green or slightly woolly above, gray woolly beneath. The inflorescence is leafy and elongate with 1/8-inch high, greenish involucres that are mostly covered with wool. There are 6 to 10 ray flowers, but they are inconspicuous. There are also 10 to 25 disk flowers. Mugwort grows in low places up to 6,000 feet. Flowers from June to October.

Tarragon (*A. dracunculus*) is an erect perennial plant that grows 20 to 60 inches tall. The leaves are linear, $1\frac{1}{2}$ to 3 inches long, entire or cleft into linear lobes. The flowering heads are small, greenish, and spreading or nodding. The involucres are up to 1/8 inch across, and the flowers are all disk, with 20 to 30 per head. Tarragon occurs on dry, disturbed places below 9,000 feet. Blooms from August to October.

Western mugwort (*A. ludoviciana*) is a perennial herb that grows 1 to 3 feet tall and has stems that are slightly white woolly above. The leaves are linear-lanceolate to elliptic in shape, 1 to 4 inches long, and entire or with a few teeth, white woolly on both surfaces or glabrous above. Inflorescence is elongate, and the involucres are about 1/8-inch high and is covered with wool. There are 5 to 12 ray flowers, but they are inconspicuous. There are 6 to 20 disk flowers. Western mugwort grows on dry, open places below 8,500 feet. Flowers from July to September.

Ecology & Ethnobotany The seeds of many species are edible raw or as flour. The seeds and peeled shoots of *A. douglasiana* and *A. ludoviciana* were eaten raw by Native Americans in California.

BALSAMROOT
Balsamorhiza   ASTERACEAE

General Description These are low perennial herbs with thick rootstalks, and the leaves are mostly basal, large, and long-petioled. The yellow flowering heads are large showy, mostly on long peduncles. Balsamroot is often confused with *Wyethia* (mule's ears), which can be found in similar habitats. However, *Wyethia* leaves lack the fuzzy gray appearance seen on the balsamroot. Arrow-leaved balsamroot (*B. sagitatta*) has the greatest history of use.

Ecology & Ethnobotany Although arrow-leaved balsamroot is considered one of the most versatile sources of food, it is not necessarily palatable. The plants contain a bitter, strongly pine-scented sap. The large taproot, root crowns, young shoots, young leafstalks and leaves, flower bud stalks, and the seeds were all eaten by various Native Americans. The larger mature leaves were often used in food preparation (i.e., wrapping).

The woody taproot of perhaps all species is edible raw or cooked. The polysaccharide inulin is the major carbohydrate found within the root. The roots can be collected throughout the year but are very difficult to dig out. In some species, the taproot may be as large as one's forearm. Cooking the roots is yet another challenge. One method we have used involves peeling the roots by pounding them to remove the bark. These were then pit cooked for 24 or more hours. When properly cooked, the roots turn brownish and sweet tasting. Another way to prepare the roots is to pit steam large quantities for a day and then mash and shape them into cakes for storage. Cooked this way, the roots were called "pash" or "kayoum."

The young shoots are edible raw or pit cooked before they emerge in early spring. The young stems and leaves can also be eaten raw or boiled as greens. The older stems are fibrous, tough, and will require some additional boiling.

The flower bud stalks are collected while the buds are still tightly closed, then peeled and eaten raw or cooked as a green vegetable. They have a slightly nutty taste.

When harvested from dried heads, the seeds can be roasted and eaten or ground into flour. The chaff is usually removed by winnowing.

PINCUSHION
Chaenactis   ASTERACEAE

General Description These wildflowers are native to western North America, especially the southwest desert of the United States. They are quite variable in appearance. They are generally aster-like in appearance with many disc florets in each head. There may be only disc florets, but sometimes there are also enlarged ray florets along the edges. They may be white to yellow or pink. Two species are of interest are Douglas' dustymaiden (*C. douglasii*) and Evermann's Pincushion (*C. evermannii*). The following key may be helpful in distinguishing the two.

1. Fruits not glandular; largest leaf blades 1-pinnately lobed; plant mat forming ----- ***C. evermannii***
1. Fruits sparsely glandular; largest leaf blades 2-pinnately lobed; plant not mat forming ----- ***C. douglasii***

Ecological and Historical Interest Douglas' dustymaiden was first described for science in 1840 by Sir William J. Hooker (1785-1865), who dedicated the specific epithet to honor the Scottish botanist David Douglas (1798-1834). Douglas discovered hundreds of new plants during his explorations of the American west.

An infusion of

the *C. douglasii* was used as a wash for chapped hands, insect bites, boils, tumors, and swellings by the Okanagon, and Thompson. A decoction of the plant was used for indigestion, coughs, and colds. A strong decoction of the plants was applied to snakebites by the Thompson, Okanagon, and Paiute.

Evermann's pincushion (formerly *C. nevadensis* var. *mainsiana*) occurs only in central Idaho (e.g., Sawtooth Mountains). It is a high altitude, mat-forming plant. The flowerheads are hairy and the leaves are irregularly pinnate (feather-like) with blunt-ended leaflets. Evermann's pincushion honors Barton Warren Evermann, a biologist and teacher.

GREEN RABBITBRUSH
Chrysothamnus viscidiflorus   ASTERACEAE

General Description Usually found in shrub-steppe type habitats in the Sawtooth Country. This is an erect, much-branched shrub. The leaves are linear to lance-shaped and flat to twisted. Flowers are aggregated into heads in rounded clusters at the stem ends. Petals are yellow. This shrub is common in dry, open areas from 4,000 to 7,500 feet and flowers from July to September.

Ecology & Ethnobotany Rabbitbrush produces a high-quality rubber called chrysil that vulcanizes easily. Because of the rubber-based compound, rabbitbrush will burn even if it is wet or green. Navajo Indians extracted a yellow dye from the flowers, while the inner bark yielded a green dye.

CHICORY
Cichorium intybus   ASTERACEAE

General Description This is a perennial herb that grows up to three feet tall with dandelion-like leaves. The blue flower heads, which can be seen from spring to fall, are composed of 15-20 or more ray flowers. The sap is milky. Chicory is a plant of waste places and is found at the lower elevations in the Sawtooth Country.

Ecology & Ethnobotany While the roasted root was used as coffee, though it is not considered a very satisfactory substitute by itself (bitter but no caffeine buzz). Many coffee producers have used chicory as a coffee additive.

The young basal leaves and flowers buds hidden at the base of the leaves are edible and best if collected from fall to spring. Because they are bitter, we found it necessary to boil them in at least 1 to 3 changes of water. When collected very young, the plants are milder when eaten raw.

THISTLE
Cirsium   ASTERACEAE

General Description Thistles are characterized as biennial or perennial herbs with alternate leaves that are lobed or cleft with spines. The red, yellow, or white heads are showy and the involucral bracts are overlapping. The native and introduced species can be found in a wide variety of habitats from the foothills to the higher elevations in the Sawtooth Country.

Ecology & Ethnobotany There seems to be little to the members of this genus, outside of the spines, which are not edible. Flowers, seeds, young leaves, and the inner parts of the stems are all edible.

The roots, crown, and inner stems can all be cooked for food. Roasting also reduces a bitter quality to the seeds.

When well-dried and de-thorned, the stems can be used as hand drills for starting fires. The stem fibers of any thistle species can be used as thread or crude cordage. To obtain the fiber, simply soak the stalks in water for a day or more to loosen them from the outer layer. The downy part of seed heads makes good insulating material and a good tinder additive.

CANADA HORSEWEED
Conyza canadensis   **ASTERACEAE**

General Description Also known as *Erigeron canadensis*. This is an annual weed which grows to about two feet tall with numerous, narrow leaves. There are numerous white flower heads. Canada horseweed is usually found growing in waste places at the lower elevations (below 6,000 feet).

Ecology & Ethnobotany A native to North America, horseweed was introduced into Europe around the mid-17th century where it became widely known for its tonic and astringent properties. A tea was made from the entire dried plant and used for gravel dropsy, diarrhea, and scalding urine. The leaves and tops of horseweed can be pounded and eaten uncooked. Native Americans used the plant in the form of a tea for leucorrhea and applied the solution to external sores in cases of gonorrhea.

HAWKSBEARD
Crepis   ASTERACEAE

General Description In general, these are perennial, tap-rooted herbs with milky juice. The leaves are alternate or all basal, and the yellow flowers are all ray. In the Sawtooth Country, the various species can be found in dry open places at lower elevations to gravelly or rocky places in alpine or subalpine areas.

Ecology & Ethnobotany The stems and leaves of *Crepis* were eaten by Native Americans. The Karok Indians in Northern California peeled the stems of *C. acuminata* before eating.

RABBITBRUSH, GOLDENBUSH
Ericameria   ASTERACEAE

General Description The species here were previously included in other genera such as *Chrysothamnus* and *Haplopappus*. These are herbs or shrubs that are often glandular. The leaves are alternate, entire to pinnatifid, and often thick. There is another rabbitbrush considered separately in this handbook and is known as green rabbitbrush (*Chrysothamnus viscidiflorus*).

Ecology & Ethnobotany The seeds and stems of some species were eaten by Native Americans.

A tea was reported to be made from the twigs of rubber rabbitbrush (*E. nauseosa*) that provided relief from chest pains, coughs, and toothaches. The leaves and stems were also boiled and the liquid was used to wash itchy areas.

Great Basin Indians were accustomed to chewing the stems of rubber rabbitbrush to extract the latex. They believed that chewing rabbitbrush relieved both hunger and thirst. The secretion obtained from the top of the roots can also be chewed as gum.

The rubber shortage of World War II stimulated research on rabbitbrush and other rubber-producing plants. Rabbitbrush produces high quality rubber called chrysil that vulcanizes easily. Because of their rubber-based compound, rabbitbrush will burn even if it is wet or green. Navajo Indians derived a yellow dye from the flowers, while the inner bark yielded a green dye.

FLEABANE
Erigeron   ASTERACEAE

General Description There are many species of fleabane in the area. They are characterized as annual, biennial, or perennial herbs with alternate or basal leaves. The flowering heads are radiate with narrow ray flowers that may be white, pink, blue, purple, or occasionally yellow. The numerous disk flowers are yellow, and the pappus is comprised of capillary bristles. The various species bloom mostly in the spring and early summer, except at the higher elevations, and can be found in a variety of habitats. The common name, fleabane, comes from the belief that these plants repelled fleas. This genus is rather difficult to work with in the field.

Ecological and Historical Interest Evermann's fleabane (*E. evermannii*) is native to the western United States. It is native to central Idaho and has also been found in western Montana. It grows at high elevations in the mountains, on steep slopes, talus outcrops, and ridges, sometime alongside whitebark pine. More precisely, the habitat is described as shifting talus slopes and dry, rocky meadows near or above timberline. The species is named for the naturalist Barton Warren Evermann (1853-1932), best known as an ichthyologist and author of Fishes of North and Middle America (1900).

WOOLLY SUNFLOWER
Eriophyllum lanatum   ASTERACEAE

General Description The yellow flowers are about 2 inches in diameter and consist of 5 to 13 ray florets (occasionally zero) and many tiny, glandular-hairy disc florets, which when mature form a semi-sphere at the plant center. The phyllaries also number from 5 to 13; they are all the same length, and covered with flattened white, woolly hairs that tend to obscure the divisions, at least towards the base. This is a rather common plant is widespread.

Ecology & Ethnobotany The seeds can be gathered, parched and then ground into flour.

CURLYCUP GUMWEED
Grindelia squarrosa   ASTERACEAE

General Description Gumweeds are biennial or short-lived perennial of waste places at low elevations. The leaves are alternate and have toothed margins. A sticky, resinous sap covers the leaves and bracts of the yellow flowers.

Ecology & Ethnobotany In general, gumweeds are considered toxic, and the toxicity appears to be dependent upon the soil in which it grows. However, many species have been used medicinally for hundreds of years.

The sticky flowers heads of *G. squarrosa* (curlycup gumweed) were used as a chewing gum substitute. The young leaves make an aromatic, bitter tea. The flower heads can be boiled in water and used as an external remedy for skin diseases, scabs, and sores. A hot poultice of the plant was used for swellings.

SUNFLOWER
Helianthus annuus   ASTERACEAE

General Description These are coarse, annual and perennial herbs, often with tall stems. Leaves are simple, the lower ones are opposite, others sometimes alternate. Flower heads are showy with bright yellow ray flowers. Involucral bracts are green and herbaceous.

Ecology & Ethnobotany The largest member of this genus in the area is *H. annuus*, is a valuable and useful plant. It has been cultivated in the United States since before Columbus. Other species of *Helianthus* may be used similarly.

The seeds may be eaten raw or roasted, then ground into meal and made into bread. The roasted shells can be used as a coffee substitute. To separate large amounts of seeds from shells, first  grind them coarsely, then stir vigorously in water. In this way the shells will float, while the seeds sink to the bottom. The tiny unopened flower buds are also edible with a flavor similar to artichokes. To reduce their bitterness boil in 2-3 changes of water. Serve with lemon and melted butter.

Sunflower oil can be extracted from the seeds for cooking, and can also be used in making soap, paints, varnishes, and candles. It is extracted by simply boiling the crushed seeds and then skimming the oil from the surface of the water. The pulp remaining after the oil is extracted also provides food for livestock.

PRICKLY LETTUCE
Lactuca serriola   ASTERACEAE

General Description This is a tall, prickly plant with alternate leaves and milky juice. The ray flowers of this species are yellow; other species may be blue, or whitish. The pappus is white to brownish. This is a rather common weed in fields and waste places in the lower elevations.

Lactuca serriola

Ecology & Ethnobotany Collected in the late fall to early spring, the plants should be boiled in a couple of changes of water to reduce the bitterness. The earlier or younger the plant is collected, the better the flavor. Because of the latex sap, raw greens can cause upset stomach if eaten in quantity. In sensitive people, the latex can cause dermatitis. These wild plants contain more Vitamin A than spinach and a good quantity of Vitamin C.

WHITE LAYIA
Layia glandulosa   ASTERACEAE

General Description This annual grows up to 24 inches tall and has leaves that are rough hairy, linear to lanceolate in shape, with the basal ones, being toothed or lobed while the upper ones are entire. The flowering heads have both ray and disk flowers present. There are about 25 to 100 disk flowers. Layia is found in sandy soil up to 7,800 feet and blooms from March to June.

Ecology & Ethnobotany The seeds of this species are edible after grinding it into flour for mush or incorporated into pinole.

TARWEED
Madia ASTERACEAE

General Description Typically, these are annuals with a tar scent of varying intensity. The leaves are narrow, usually opposite below and alternate above. The flower heads are comprised of inconspicuous yellow ray flowers. In the Sawtooth Country, at least 3 species of tarweed can be found at the moderate elevations in open, grassy, or vernally moist areas.

Ecology & Ethnobotany The seeds of all tarweed seeds were collected and stored until needed. Seeds were often used in making pinole by Native Americans. Some tribes pulverized tarweed seeds and ate them dry. The roots of some species were also eaten.

The scalded seeds also yield a nutritious oil. The oil was used like olive oil before olive was readily available in this country.

PINEAPPLE WEED
Matricaria discoidea ASTERACEAE

General Description This is an annual herb with a branched habit. Leaves are alternate and pinnately lobed or divided. The small, terminally arranged flower heads are composed of disk or ray flowers. This is an introduced plant with a circumboreal in distribution. It can be easily recognized by its pineapple-like smell when crush between the fingers.

Ecology & Ethnobotany A delicious tea can be made from the dried flowers of the plant. The leaves are edible, but bitter. The medicinal uses of pineapple weed are identical to that of chamomile (*Anthemis*). Used as a tea it is a carminative, antispasmodic, and mild sedative.

NODDING SILVERPUFF
Microseris nutans   ASTERACEAE

General Description This is a taprooted perennial with milky juice. The leaves are basal, and the flower heads are always ligulate and yellow. Look for the plant in open and moist habitats of the montane and subalpine zones.

Ecology & Ethnobotany The slender roots of nodding silverpuff are apparently edible raw.

GOLDENROD
Solidago   ASTERACEAE

General Description The various species of goldenrod are perennial herbs with fibrous roots. The leaves are alternate, simple, and either tooth or entire. The heads are made up of yellow ray flowers. In the Sawtooth Country, goldenrods may be found in dry to moist habitats from the foothills to timberline, often in dense patches.

Ecology & Ethnobotany Young leaves can be prepared as potherbs or added to soups. Depending on habitat, age, and personal preference, their palatability is quite variable. The dried leaves and dried, fully expanded flowers can be used to make a tea. The seeds can be used to thicken stews. Large amounts of the raw herbage should be avoided as it may be toxic.

SOW THISTLE
Sonchus   ASTERACEAE

General Description The species are introduced from Europe and are weedy annuals with alternate leaves that are entire to pinnately divided. The leaf bases have eared-shaped lobes and the margins are prickly. The flower heads are composed of entirely yellow ray flowers. The pappus is bristly. In the Sawtooth Country, the species can be found in waste places in lower elevations.

Ecology & Ethnobotany The young plants of all species can be prepared as a potherb. As they get older they become increasingly bitter. We found that boiling them in at least two changes of water makes them a little more palatable. Since the plants have an abundance of soluble vitamins and minerals, use only a minimum amount of water and boil briefly.

DANDELION
Taraxacum officinale   ASTERACEAE

General Description Dandelions need very little introduction. All species are tap-rooted perennials with milky juice and leaves that form a dense, basal rosette. The solitary flower head is composed of bright yellow ray flowers. They are found in a variety of habitats up to the alpine zone.

Ecology & Ethnobotany Every part of dandelion is edible. The young leaves may be eaten raw or cooked like spinach. The older leaves are also edible, but we find it is better to boil the older leaves in 1 or 2 changes of water to eliminate the bitterness that comes with age. The plants are high in Vitamins A and C, a good source of B complex, and iron, calcium, phosphorous, and potassium. The roots can also be eaten raw, or boiled as a vegetable, baked as potatoes, or added to soups and stews. The roasted root can be used as a substitute for coffee, but it lacks the caffeine buzz. The flower buds can be pickled and added to meals such as omelets.

> *There is only one safe way to use wild plants for food: you must know, positively and beyond doubt, that the plant you propose to eat is edible!* **It's a question of being dead certain or of being dead, period.** *If you cannot identify a plant as edible beyond all doubt, do not eat it!*

GOLDENWEED
Tonestus   ASTERACEAE

General Description These are taprooted or rhizomatous perennials, glandular or hairy. The leaves are alternate or all basal and entire. The flowering heads are usually 1 per stem and the rays are yellow. Two species found in the Sawtooth Country: Idaho Goldenweed (*T. aberrans*) and Lyall's Goldenweed (*T. lyallii*).

1. Rays absent ----- ***T. aberrans***
1. Rays usually present ----- ***T. lyallii***

Ecological and Historical Interest Based on Flora of North America, Idaho goldenweed is now referred to as *Triniteurybia aberrans* and is only found in the Sawtooth Mountains of Idaho and the Bitterroot Range of Montana. It is considered to be a sensitive species. *Tonestus* is a meaningless anagram of *Stenotus*.

SALSIFY, GOAT'S BEARD
Tragopogon ASTERACEAE

General Description The two species in the Sawtooth Country are introduced, tap-rooted, biennial herbs with milky juice. The leaves are alternate, entire, sessile and clasping at the base and taper to a long point. The flower heads are solitary and composed of pale yellow or purple ray flowers. The heads open early in the day, close about noon and remain closed on cloudy, rainy days. They are found in many habitats

Ecology & Ethnobotany The fleshy roots of the purple flowered *T. porrifolius* (salsify) can be eaten raw or after cooking. The flavor resembles that of an oyster, an acquired taste! The yellow flowered species, *T. dubius* (yellow salsify) and *T. pratensis* (meadow salsify), are also edible, but are somewhat smaller, more fibrous, and tough. Salsify root has been cultivated for over 2,000 years in the Mediterranean. The young leaves and stems of all species can be eaten after boiling until tender. The coagulated sap can be used as chewing gum and as a remedy for indigestion.

WYETHIA
Wyethia   ASTERACEAE

General Description These are stout perennial, simple stemmed herbs with large, erect, alternate leaves. The ray flowers are long and yellow or white. Heads are usually solitary. All species have leaves on the stems distinguishing them from *Balsamorhiza*, which only has leaves at the base.

The plant's scientific name (*Wyethia*) was for Nathaniel Wyeth, an early fur trader. He established

Fort Hall, a trading post near present-day Pocatello, Idaho. In 1834 Wyeth accompanied a botanist on an expedition across the country, and this plant was collected and described. *W. helianthoides* is similar in appearance to *W. amplexicaulis*, which has yellow-rayed flowers. Both seem to hybridize where they overlap.

Ecology & Ethnobotany The roots are edible after long cooking. Seeds are edible too and resemble sunflower (*Helianthus*) in taste. **The leaves, if nothing else, are supposedly poisonous.**

BARBERRY, OREGON-GRAPE
Berberis aquifolium   BERBERIDACEAE

General Description These are shrubs with pinnately compound, evergreen leaves. The leaflets have spiny margins, and the yellow flowers are in 3 whorls that are interpreted as bracts, sepals, and petals. The flowers have six or more stamens that split open by two hinged valves to splatter pollen over insects as they crawl by. Use a hand lens to view the unique anthers, which open to release pollen by a pressure-controlled, flap-like valve, instead of splitting down the side. The fruits are blue to purple in color and have a waxy covering. Another species may also be encountered and the following key may help in distinguishing the species.

1. Stems to 6 feet tall; leaflets averaging twice as long as wide or more, not papillose (with small rounded bumps) beneath ----- **M aquifolium**
1. Stems to 3 feet tall; leaflets averaging less than twice as long as wide, minutely papillose beneath ----- **M. repens**

Ecology & Ethnobotany The blue berries are edible raw or can be dried for future use or added to soups to improve flavor. The plants contain berberine, a bitter alkaloid that gives roots their distinctive yellow color and usefulness as a digestive tonic. Berberine stimulates the involuntary muscles and possesses anti-pyretic, laxative, and anti-bacterial qualities. A yellow dye can be obtained by boiling bark and roots.

ALDER
Alnus   BETULACEAE

General Description The species are small trees or shrubs with smooth, reddish or gray-brown bark. Leaves are egg-shaped and have serrate edges. The male catkins are grouped near the end of branches and drop off after pollen is shed. The female catkin is cone-like and persistent. Fruits are flattened achenes with lateral wings or just a membranous border. These plants are usually associated with riparian and wetland sites at low to mid elevations. Two species include: gray alder (*A. incana*) and green alder (*A. viridis*).

1. Catkins appear on previous year's growth before leaves unfurl; female catkins ("cones") borne on peduncles much shorter than length of body ----- ***A. incana***
1. Catkins and leaves appearing simultaneously on twigs of current season; female catkins with peduncles longer than body of catkin ----- ***A. viridis***

   Ecology & Ethnobotany Since alders usually grow in the vicinity of free-flowing water, it is considered a botanical indicator of water. The edible catkins are high in protein, but generally don't taste very good. The catkins are more tolerable if they are nibbled raw, added to soups, or dried and powdered and used as a spice. The inner bark is palatable only for a short time in the spring when it is less bitter. A patch of bark is removed from the tree and the tissue scraped off and eaten fresh or dried in cakes. Alder is valued for its hardwood and is useful for open fires as it does not readily spark. It is used widely by Native Americans for woodworking, including dishes, spoons, and platters. The wood is also used for making fire drill sets.

BIRCH
Betula   BETULACEAE

General Description Two species can be found in the Sawtooth Country: water birch (*B. occidentalis*) and bog birch (*B. pumila*). They are described as deciduous trees or shrubs with simple, alternate, and sharply toothed leaves. Birches can be found along streams, and in wet meadows and bogs from the foothills to upper montane zone. The following key may help in distinguishing the species.

1. Margins of leaf blade crenate to blunt-dentate; shrubs with close bark ----- ***B. pumila***
1. Leaf blade margins simply or doubly serrate to dentate, teeth obtuse to relatively sharp; trees and shrubs with close or exfoliating bark ----- ***B. occidentalis***

Ecology & Ethnobotany Young birch leaves can be added to salads. The inner bark can be dried and ground into flour, and the twigs can be steeped in hot water for a tea. The juice of birch leaves makes a good mouthwash.

Birch contains a significant amount of methyl salicylate and is often used in teas for headaches and rheumatic pain. Birch is highly regarded as a medicinal plant in Russia and Siberia for treating arthritis.

Since it burns even when wet, birch bark makes a good tinder. The sap, collected in much the same manner as maple, was sometimes made into syrup or vinegar. The best time for tapping is early spring, before the leaves unfurl.

ARCTIC ALPINE FORGET-ME-NOT
Eritrichium nanum   BORAGINACEAE

General Description These are dwarf, cushion-like perennial herbs. The leaves are densely crowded on numerous short shoots, and more or less hairy. The blue flowers with a yellow center occur in clusters. There are 1-4 nutlets.

Ecological and Historical Interest The showy blue flowers of arctic alpine forget-me-not are pollinated by a wide array of insects ranging from syrphids to bees. Seeing these beautiful, blue alpine plants is a special treat for hikers and climbers. It is the just award for reaching the "top of the mountain."

BLUEBELLS, LUNGWORT
Mertensia ciliata   BORAGINACEAE

General Description This is a perennial herb with succulent, alternate, and entire leaves. The blue flowers are funnel-form or trumpet-shaped. The genus is named after the German botanist, F.K. Mertens. The common name, lungwort, comes from a European species with spotted leaves, believed to be a remedy for lung disease. Four other species occur in the Sawtooth Country and can be used in similar ways.

Ecology & Ethnobotany Bluebells are often overlooked in many edible plant guides. The flowers can be nibbled upon raw or added to salads. Since the leaves are a bit hairy, we found them better when chopped up and added to soups. Bluebells may contain alkaloids and other constituents that can be toxic if consumed in large quantities.

ALYSSUM, MADWORT
Alyssum **BRASSICACEAE**

General Description These are small annual plants with foliage that appears dull gray due to the presence of dense, star-shaped hairs. The leaves are alternate and simple, and the flowers are short-stalked on the terminal portion of the stems. The flowers have light yellow petals that quickly fade to white. Fruits are egg-shaped or round in outline and have winged margins and a short style.

Ecology & Ethnobotany Both species are Eurasian weeds and are widespread in distribution throughout most of the United States. The genus name is from the Greek *a* (without) and *lyssa* (rabies), to the supposed cure for rabies. The small leaves of pale madwort are mild tasting and can be eaten raw.

ROCKCRESS
Arabis & Boechera BRASSICACEAE

General Description These are biennial or perennial herbs with stellate hairs. Flowers occur in racemes and are usually white to purple in color. The fruits are linear siliques, usually flattened parallel to the partition. The species are found in a variety of habitats at various elevations.

Ecology & Ethnobotany Though *Arabis* was traditionally recognized as a large genus with many Old World and New World members, recent studies of these species using genetic data suggest that they are not closely related, so *Arabis* has been split into two separate genera. The Old World members all remain in the genus *Arabis*, whereas most of the New World members have been moved into the genus *Boechera*, with

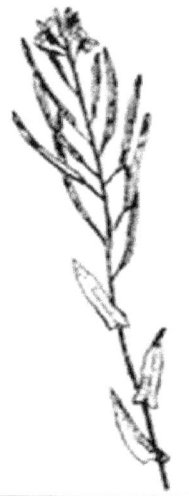

only a few remaining in *Arabis*. From an edibility perspective, they are similar.

The author has enjoyed forging on the first year's rosettes; a little bitter tasting but refreshing.

WINTERCRESS
Barbarea orthoceras   BRASSICACEAE

General Description This is an erect glabrous perennial with stout angled stems. The basal leaves are 1-4 inches long, pinnatifid with a large terminal leaflet. The leaves on the main stem are pinnatifid or lobed. Flowers occur in a dense raceme and are pale yellow in color. The mature fruits are 1-2 inches long. In the Sawtooth Country, look for wintercress in moist places and along stream banks between 2,500 and 11,000 feet.

Ecology & Ethnobotany The young plants are best when gathered in early spring. They are good in salads or as a potherb. Older leaves can still be eaten by parboiling them to remove the bitterness. The flowers clusters, even when opened, can be substituted for broccoli.

BLACK MUSTARD
Brassica nigra   BRASSICACEAE

General Description *Brassica's* are large annuals with showy yellow flowers. The pods are round or 4-sided in cross section, with a conspicuous beak. The species may be encountered in waste places and fields at lower elevations.

Ecology & Ethnobotany All *Brassica* species are edible. When young, the plants can be eaten raw or cooked. As it matures, the leaves can be cooked for a potherb; they require fairly lengthy cooking at this point.

Mustard seeds are best known for flavoring because of their sharp taste. Flowers and flower buds are palatable. The pollen is a good protein source.

FALSE FLAX
Camelina sativa   BRASSICACEAE

General Description This is an introduced genus of annual or biennial herbs with all but one species, *C sativa*, occurring wild or as a weed. The leaves are entire or toothed, those on the stem sessile and clasping at the base. The yellow flowers are small and occur in rather long racemes.

Ecology & Ethnobotany *C. sativa* has been cultivated since the Neolithic Age for the fibers of its stem and the edible oil contained in the seeds. It is said that an oil similar to linseed oil.

SHEPHERD'S PURSE
Capsella bursa-pastoris   BRASSICACEAE

General Description This is an erect annual growing up to 20 inches tall. It has basal leaves that are dissected. The upper stem leaves become reduced in size and are more entire, lanceolate in shape and sessile with clasping base. The flowers are small, white in color. The fruits are heart-shaped or purse-shaped. Along the

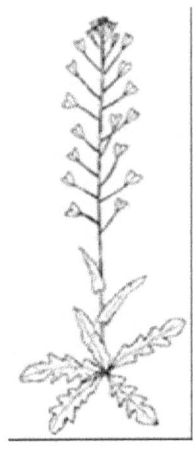

trail this plant may be found on dry soils and disturbed areas at the lower elevations.

Ecology & Ethnobotany Shepherd's purse has been used as food for thousands of years. The young leaves are edible raw or cooked. They are delicious when blanched. The plant is extremely high in Vitamin K, the blood clotting vitamin. Seeds can be ground into meal, or they can be used for flavoring other foods. Roots have been candied by boiling in syrup. Fresh or dried, they can be used in place of ginger. The pods are also useful as food.

TOOTHWORT, BITTERCRESS
Cardamine   BRASSICACEAE

General Description This large genus contains more than 150 species of annuals and perennials. The genus grows worldwide in diverse habitats. The flowers are white or purple and the pods are elongate and flattened. Three species may be found in the Sawtooth Country. Several species were previously included in the genus *Dentaria*.

Ecology & Ethnobotany The plant is acceptable in salads. However, we suggest cooking them in at least a change of water to improve the taste. The various species have various bitter tastes. In a pinch, they are probably all useful. Some plants were reputed to have medicinal qualities (treatment of heart or stomach ailments).

TANSYMUSTARD
Descurainia BRASSICACEAE

General Description Tansymustards are annual or biennial herbs with leaves that are 1-3 times pinnately divided. The foliage is covered with simple, branched, or short gland-tipped hairs. The flowers are cream-colored or light yellow and the pods are long, narrow, 3-sided to nearly round in cross section. The species in the Sawtooth Country are weedy and occur in disturbed soils at the lower elevations.

Ecology & Ethnobotany The young green parts of these plants can be steamed for about half an hour; they are then eaten or dried for future use. Parboiling helps to remove the bitter taste. The seeds were parched by tossing in a basket with hot stones or live coals, then ground into a fine flour and made into mush. Because of its peppery taste, the mush was often mixed with the flour of other seeds to make it more palatable. This species is known to cause poisoning in cattle.

WALLFLOWER
Erysimum BRASSICACEAE

General Description Wallflowers are annual, biennial, or perennial herbs that are often tap-rooted. The herbage is covered with closely appressed forked hairs and yellow or orange flowers are showy. The linear pods are 4-sided in cross section with a small beak. Four species may be found in the Sawtooth Country.

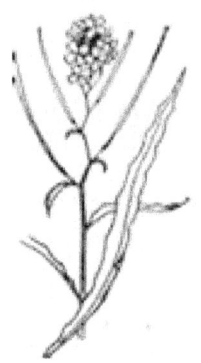

Ecology & Ethnobotany
As with most mustards, the can be eaten, but are not choice. Wallflowers were once used as a poultice. *Erysio* means to draw out, as in drawing out pain or causing blisters.

PEPPERGRASS, PEPPERWEED
Lepidium   BRASSICACEAE

General Description The various species in the Sawtooth Country are non-native annual or biennial herbs with simple or 1-3 pinnately divided leaves that are alternate or basal. The flowers are white, yellow, or greenish, and the pods are flattened at right angles to the partition that separates the seed chambers. Look for these peppergrasses in dry, open or vernally moist areas at low to mid elevations.

Ecology & Ethnobotany
The young stems and leaves may be eaten raw or dried for future use. The plants contain Vitamins A and C, iron, and protein. The seed pods and seeds can be used as a flavoring.

WATERCRESS
Nasturtium officinale   BRASSICACEAE

General Description This is an aquatic perennial that is slightly juicy or succulent. Leaves are pinnate compound into 3 to 11 ovate leaflets. The flowers are white or yellow, and the fruit is a curved pod. Watercress is common in quiet streams or on wet banks below 8,000 feet. The small white flowers, succulent foliage, and aquatic habitat will immediately identify watercress. Flowers from March to November.

Ecology & Ethnobotany The peppery-tasting plants were eaten raw or cooked as a potherb. A good source of vitamins, watercress is listed as efficient in preventing scurvy. Watercress contains significant amounts of iron, calcium and folic acid, in addition to vitamins A and C. In some regions watercress is regarded as a weed, in other regions as an aquatic vegetable or herb. Where watercress is grown in the presence of animal waste, it can be a haven for parasites such as the liver fluke.

As noted, the herbage of watercress is edible if the waters in which they grow are not polluted. However, finding unpolluted water may be difficult. One suggestion would be to soak the fresh greens in a disinfectant, or treat the water with water purification tablets, or a tablespoon of bleach in a quart of water. Then rinse the greens well in potable water to remove the chemicals.

YELLOWCRESS
Rorippa   BRASSICACEAE

General Description These are tap-rooted annuals or rhizomatous perennials with simple or pinnately divided leaves. The flowers are yellow or white; the pods are elliptical to linear and 3 sided to slightly compressed. In the Sawtooth Country, the various species occur in moist, wet, or aquatic habitats up into the middle elevations. *Rorippa curvipes* and *R. curvisiliqua* are relatively common and widespread species.

Ecology & Ethnobotany The leaves of most species are edible raw or after cooking. The plants are rich in iron, copper, calcium, sulfur, and magnesium. They also contain substantial quantities of Vitamins A, B, B2, and C.

TUMBLEMUSTARD
Sisymbrium   BRASSICACEAE

General Description These are annual, biennial or perennial herbs. The small flowers are yellow or white, and the fruits are linear. The three species occur in the area introduced from Europe and are widespread throughout the United States. The various species are found in waste places and disturbed habitats at low elevations.

Ecology & Ethnobotany The seeds of *S. officinale* can be parched and then ground into a flour. The plants also make fine potherbs. As with other mustards, it is best to cook the plants in a couple of changes of water.

PRINCE'S PLUME
Stanleya   BRASSICACEAE

General Description The species along in the Sawtooth Country are annual or perennial herbs. The flowers are in elongated racemes. The fruits (siliques) are borne on a long stipe. The plants are usually found in sagebrush habitats at low elevations. The genus is named for Lord Edward Stanley, a British ornithologist who lived from 1775 to 1851.

Ecology & Ethnobotany The tender leaves and stems of all species can be prepared in much the same way as cabbage. They are bitter but boiling in several changes of water remove some of the astringent properties. The seeds can be collected, parched, and then ground into a flour. They can be eaten as a mush or used in making breads.

COMMON FRINGE-POD
Thysanocarpus curvipes   BRASSICACEAE

General Description This is a slender branched annual with stem leaves and basal leaves arranged in a rosette. The flowers are purplish, and the circular, flattened pods are surrounded by a flat nearly circular wing. The species occurs in open areas at low to mid elevations.

Ecology & Ethnobotany Fringe-pod seeds are edible after parching and being ground into flour. A tea

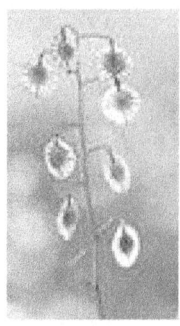

made from the plant is said to cure stomach-ache; a drink made from the leaf can be used to relieve colic.

WATERSHIELD
Brasenia schreberi   CABOMBACEAE

General Description Watershield is anchored to muddy substrates by slender rootstocks. All exposed portions of the plant are covered with a gelatinous sheath. The leaves are nearly round and arise near the tops of the stems. The flowers have purplish petals and sepals.

Ecology & Ethnobotany The starchy rootstalk of watershield can be peeled, boiled, and eaten, or dried and stored or ground into flour. The unexpanded young leaves and leaf stems can also be eaten in a salad, slime and all. In Japan, it is an ingredient of miso-shiru (miso soup). In China, may be fried as food. When cooked with crucian carp and bean curd, the soup is said to be fragrant and tasty. The rootstalks were used to cure dysentery and stomach-ache. The plants are reported to have antibacterial, antialgal activity and to be allelopathic to lettuce seedlings.

SPINYSTAR
Escobaria vivipara   CACTACEAE

General Description The cacti in this genus are ball-shaped (about $2\frac{1}{2}$ inches in diameter) and have greenish-white to deep red or purple flowers that are borne at the tip of the stem. Imagine a group of whitish tennis balls with spines. They are usually found at the lower elevations in the foothills. These species were once classified in the genus *Corypantha*.

Ecology & Ethnobotany Like other species of cactus (e.g., *Opuntia*, etc.), the ripe fruits are edible. It is usually best to boil before eating. Spinystar is also found growing in Canada and more northern areas where other cacti do not. It has been suggested that these cacti are more "freeze tolerant" than its cousins. However, too much water appears to rot the plants, probably why it is found growing in the drier lower elevations.

BEAVERTAIL, PRICKLY PEAR
Opuntia   CACTACEAE

General Description Prickly pear need little introduction. These are succulent herbs with fibrous roots and the stems that are flat. The leaves when present are small, fleshy, and awl-shaped. Many *Opuntia* species have glochids - minute, nearly invisible barbed hairs that grow in clusters in areoles. They easily become embedded in skin or clothing, and, because of their light tan or yellowish color and barbed surface, are almost impossible to remove. These cacti can occur in dry soils at various elevations.

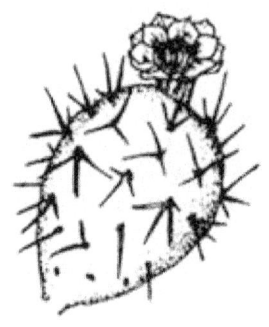

Ecology & Ethnobotany The fruit of *Opuntias*, called tuna, is edible, although it has to be peeled carefully to remove the small spines on the outer skin before consumption. It is often used to make candies and jelly. The young stem segments, called nopales, are also edible. The gel-like liquid of a prickly pear cactus can be used like a conditioner. Prickly pears also have medicinal uses. They are said to control blood sugar, cure acne, and soothe skin, and can also be used as arthritis medicine and eye drops.

HAREBELL, BLUEBELL
Campanula   CAMPANULACEAE

General Description These are perennial herbs from a rhizome. The blue (occasionally white) flowers are tubular-, bell- or cup-shaped. The genus name is from the Latin "bell," and the common name, harebell, may allude to an association with witches, who were believed to transform themselves into hares, porters of bad luck when they crossed a person's path. The species can be found in open, dry, or rocky areas from low elevations to above timberline.

Ecology & Ethnobotany The leaves and shoots of at least *C. rotundifolia* (bluebell bellflower) can be used in salads or cooked as a potherb. The roots can also be boiled and eaten and have a nut-like taste. *C. rapunculoides* (rampion bellflower) is also edible.

ROCKY MOUNTAIN BEEPLANT
Cleome serrulata   CAPPARACEAE

General Description This is an erect, showy plant up to 40 inches tall with alternate leaves divided into three lance-shaped, entire leaflets. The reddish-purple to pink flowers are arranged in a dense, narrow, terminal inflorescence. The petals are separate while the sepals are united. The fruits are long-stalked, pendulous capsules, linear to lance-shaped in outline. Bee plant is found in disturbed areas (i.e., roadsides, railroad right-of-way) in the Sawtooth Country.

Ecology & Ethnobotany An important food for many western Native Americans, bee plant was extensively used as a potherb. The young tender shoots and leaves, and flowers are preferred. The plant has an unpleasant odor, especially when older, and a pungent taste much like the mustards. We found it necessary to cook the plants in at least two changes of water to remove the bitter taste. The seeds can also be collected and ground into flour.

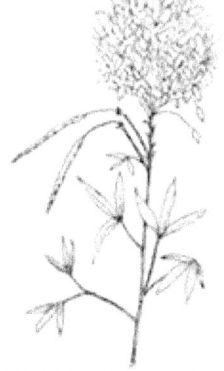

As a dye, plants are collected in quantity and boiled down for several hours until a thick, fluid residue is produced. The water is then drained off and the plants allowed to dry and harden into cakes. When black dye or paint is needed, a piece of the cake is soaked in hot water.

HONEYSUCKLE, TWINBERRY
Lonicera   CAPRIFOLIACEAE

General Description The species here are shrubs or woody vines with entire and opposite leaves. In the Sawtooth Country, they can be found in a variety of habitats from the foothills up to the alpine zone. The flowers are quite fragrant and we've usually smelled the flowers before actually seeing it.

Ecology & Ethnobotany The berries of this species are edible and can be eaten raw or dried for future use. The long stems of honeysuckle can be used as basket foundation material. You can also peel and split the hairy stems and use as wrapping material for coiled baskets.

ELDERBERRY
Sambucus nigra   CAPRIFOLIACEAE

General Description Elderberries are shrubs with pithy stems. The two species in the Sawtooth Country have large, compound leaves with serrated leaflets. The white flowers are arranged in dense clusters. The fruits may be red (*S. racemosa*) or blue-black (*S. nigra*). Elderberries can be found in open areas, hillsides, and riparian habitats in the montane zone.

Ecology & Ethnobotany The blue or black colored berries of *S. nigra* are edible raw, or they can be made into excellent jams, jellies, and wine. They can also be dried and stored for future use. The seeds contain hydrocyanic acid, and if eaten in quantity can cause diarrhea and nausea. It is best to cook the berries or strain the seeds before use. ***The red-berried species should be avoided as it is best described as poisonous.***

SNOWBERRY
Symphoricarpos   CAPRIFOLIACEAE

General Description Snowberries are erect shrubs with elliptical to egg-shaped leaves. Flowers are white to pink and bell-shaped, accompanied by two small bracts. The fruits are berry-like and white. Four species are found in dry soils at various elevations.

Ecology & Ethnobotany The white, tasteless berries are edible raw or cooked, and are said to be emetic and cathartic in large amounts. Saponins are found in the leaves and can be

used as a natural cleaning agent. The new twigs are flexible and can be used in cordage and basketry.

LARGE MOUSE EARS
Cerastium glomeratum   CARYOPHYLLACEAE

General Description Of the six species, this is probably the most likely one a hiker will likely to encounter in the Sawtooth Country. This species of chickweed is an annual with opposite, narrow, ovate leaves. The herbage is usually hairy and sticky. The

flowers are white and the petals are deeply lobed at the tip. The fruit is a cylindrical capsule, often slightly curved at maturity. This is a common plant of waste places at lower elevations.

Ecology & Ethnobotany *Cerastium* is frequently confused with *Stellaria media* (chickweed), but to the general forager there is no danger. The tender leaves and stems of most *Cerastium* can be added to a salad, but we found they are better if boiled first and served as greens.

TUBERED STARWORT
Pseudostellaria jamesiana   CARYOPHYLLACEAE

General Description This is a weak-stemmed, glandular perennial with lanceolate leaves and many few-flowered cymes of white flowers that have slightly 2-lobed petals. This species is found about meadows and damp places and flowers May to July. *Pseudostellaria* refers to starwort's resemblance to the genus *Stellaria*.

Ecology & Ethnobotany The tuber-like swellings of this species can be eaten raw or dried in the sun. They have a thin, light brown rind, and a tender rather mealy texture inside, similar to a potato.

BOUNCING BET
Saponaria officinalis   CARYOPHYLLACEAE

General Description This an erect perennial herb with sessile or nearly sessile leaves. The flowers are showy, usually pale pink. Bouncing bet can be found along roadsides, disturbed areas, and waste places at the lower elevations. The plant has escaped from cultivation. The genus name is Latin for soap, since the juice of the plant lathers with water.

Ecology & Ethnobotany The plant contains saponins and will irritate the digestive tract if eaten. The crushed green plant and roots can be used as a soap substitute.

PURPLE SAND SPURRY
Spergularia rubra   CARYOPHYLLACEAE

General Description *Spergularia* can be distinguished from other similar small Caryophyllacious plants by presence of stipules - small appendages at bases of leaves.

Ecology & Ethnobotany The tiny seeds are sometimes eaten, but contain saponin. They were gathered, ground up, and mixed with flour to make bread.

COMMON CHICKWEED
Stellaria media   CARYOPHYLLACEAE

General Description This is a slender, weak-stemmed annual with trailing stems. The leaves are opposite and the flowers are small and white. Looking closely at the petals, at first it appears as though there are 10 petals, but in actuality there are only 5; the petals are what we call bifid. These is a rather common plant in moist shady areas along the trail, especially parks and lawns. The other species occur higher in the mountains and in natural habitats.

Ecology & Ethnobotany There are 6 species *Stellaria* in the Sawtooth Country. While the uses of other *Stellaria* are unknown, the young shoots of this species have been used as salad herbs or potherbs if cooked like spinach. Although it is edible raw, we prefer to boil for a few minutes before eating. Since the plants are usually quite small and only the youngest parts are good, chickweed can be tedious to collect. The greens are low in calories and packed with copper, iron, phosphorus, calcium, potassium, and Vitamin C - valued in the prevention and treatment of scurvy.

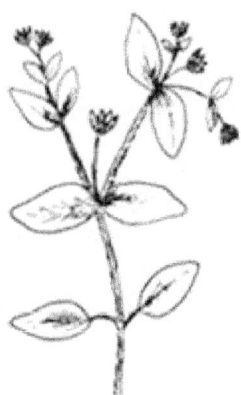

MOUNTAIN LOVER
Paxistima myrsinites   CELASTRACEAE

General Description This is a low, dense, evergreen shrub. The thick, leathery leaves are opposite, oval to elliptic in shape, and the toothed margins are slightly rolled under. The small flowers are maroon in color. The plant is found in coniferous forest, rocky openings, and dry mountain slopes from low to mid elevations. The genus is often spelled *Pachistima*.

Ecology & Ethnobotany The fruits were eaten by some Native Americans in California, while other sources indicate that the plant is inedible. ***Best to treat the fruit as inedible.*** Additionally, they used the boiled leaves to poultice pain or inflammation.

COON'S TAIL
Ceratophyllum demersum   CERATOPHYLLACEAE

General Description This is a rootless, submersed or free-floating aquatic plant with slender, lax, and much branched stems. The sessile leaves are in whorls of 5-12, and the blades are dissected into linear, filamentous segments whose shape varies with the position on the plant. The minute flowers have no petals, and are borne in the axils of the leaves. Coon's tail is a common plant in standing or slowly flowing water of rivers, sloughs, and ponds at low to mid elevations.

Ecology & Ethnobotany Native Americans used coon's tail to make a soothing lotion that was used on sore or inflamed skin.

SALTBUSH
Atriplex   CHENOPODIACEAE

    General Description These are annual or perennial herbs or shrubs with alternate leaves, and glabrous or scaly herbage. The flowers are unisexual, and individual plants have one or both sexes. The various species are found at the lower elevations in valleys, disturbed areas, or in dry, alkaline soils.

    Ecology & Ethnobotany The many uses for these plants include food to medicine and dyes, as well as soap and spice. The young leaves of many species can be cooked and eaten as greens and have a very distinct salty taste. We've often added them to otherwise bland foods to make our wild meals less boring. Add the leaves to meats while cooking will help spice them up. The seeds were parched, ground into flour, and made into mush. They can also be soaked in water for a few minutes to make a rather pleasant tasting drink. The Navajo used the flowers to make puddings. The ashes of *A. canescens* (fourwing saltbush) make a good substitute for baking soda.

    The leaves and roots of many species were used as soap. They were rubbed in water for lather and used in washing clothing and baskets. Many Native Americans also carved arrowheads from the wood for use as weapons and for hunting. The seeds of some species were also used in making a black dye.

COMMON KOCHIA
Bassia scoparia CHENOPODIACEAE

General Description Formerly in the genus *Kochia*. Of the three along the trail, common kochia is the more likely species to be encountered. Common kochia is a bushy annual with stems up to three feet tall. The leaves are alternate, narrowly lance-shaped and tapered at both ends. The herbage may or may not be covered with hairs. Flowers are solitary or in clusters in spikes. The species is common in open, disturbed habitats at low elevations.

Ecology & Ethnobotany Common kochia is native to Europe and was introduced into the United States as an ornamental. It has since escaped and has become well established. In Asia, Japan, and China, common kochia was cultivated for its seeds. The tips of the young shoots can be prepared as potherbs. The seeds can be eaten raw or cooked, or ground into meal and used in bread making.

GOOSEFOOT
Chenopodium CHENOPODIACEAE

General Description The species in this genus are annuals with mealy or glandular foliage. Leaves are alternate and entire, toothed, or lobed. Flowers are borne in dense clusters in the leaf axils or terminal inflorescence. There are 2 to 5 green or red sepals.

Ecology & Ethnobotany Leaves, tops, and seeds of all species can be used as an emergency or basic food and are quite tasty and nutritious. High in

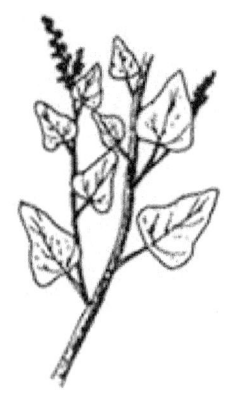

protein, the greens are a good source of Vitamins A and C, iron, potassium, and are extremely rich in calcium. Since it does not become bitter with age, both young and old plants can be used. Leaves may be used raw in salads or boiled in water like spinach. The water can be saved and used as a yellow dye. The leaves were also eaten to treat stomachaches and prevent scurvy. A leaf poultice was used on burns. The flower buds and flowers can be used as potherbs. A single plant can produce up to 70,000 seeds. Seeds can be ground as flour for use in bread or cooked as mush. Seeds can also be eaten without grinding or incorporated into pinole (flour made from a mixture of seeds of small plants). The seeds contain about 15% protein and 55% carbohydrates, more than is found in corn. The seeds can also be used as a coffee substitute.

JERUSALEM OAK GOOSEFOOT
Dysphania botrys   CHENOPODIACEAE

General Description Formerly known as *Chenopodium botrys* and as *Ambrosia mexicana*. The genus *Dysphania* is known as the glandular goosefoots.

Jerusalem oak goosefoot is native to the Mediterranean region.

Ecology & Ethnobotany The leaves and seeds are more or less edible. Leaves can be cooked or raw leaves should only be eaten in small quantities. The seeds can be ground into a meal and used with flour in making bread. The seed is small and fiddly, it should be soaked in water overnight and thoroughly rinsed before it is used in order to remove any saponins. The leaves are a tea substitute.

A gold or green dye can be obtained from the whole plant. Additionally, the dried plant is a moth repellent. The whole plant is very aromatic and is used as a scent in pillows, bags, and baskets.

NUTTALL'S POVERTYWEED
Monolepis nuttalliana CHENOPODIACEAE

General Description Povertyweed is a low growing winter annual with prostrate or ascending stems. The leaves are somewhat succulent and lance-shaped, broadened and lobed at the base. Flowers are borne in dense clusters at the leaf bases and the solitary sepal is reddish in color. The seeds are dark brown. The plant can be found in open disturbed habitats at the lower elevations.

Ecology & Ethnobotany The above ground parts of povertyweed may be eaten as a potherb. The seeds are also edible. Another species (*M. spathulata*) may also be encountered along the trail, but its uses are unknown.

PRICKLY RUSSIAN THISTLE
Salsola tragus   CHENOPODIACEAE

General Description This is not a true thistle (*Cirsium*), but a many branched annual with purplish striped stems up to three feet tall in a rounded form. The lower leaves are threadlike; the upper leaves are awl-like and spine-tipped. The plant may or may not be hairy. When mature, the whole plant becomes rigid, breaks off at ground level, and becomes one of the "tumbleweeds" that blows across the open plain. Flowers are solitary in the leaf axils and are subtended by spiny bracts. Russian thistle is common in open, disturbed habitats, particularly around agricultural areas at low elevations. It was introduced to the United States from Europe. It is considered a noxious weed because of its distributional pattern and spines.

Ecology & Ethnobotany This unsavory looking plant is edible. The young parts of the plant may be boiled and eaten as a potherb or chopped raw into a salad. On older plants, clip the tender branch tips that are green. We find the taste of the plant greatly improves when cooked in butter and lemon. In Europe, the ashes of the plant were once used in the production of carbonate of soda known as Barilla. Warning The older parts of the plants contain significant quantities of nitrates and oxalates and may be toxic if eaten in quantity.

ST. JOHN'S WORT
Hypericum CLUSIACEAE

General Description These species have yellow flowers, and small, translucent glands on the leaves and petals. The species can be found in moist areas at various elevations. Two species may be encountered in the Sawtooth Country: the creeping St. John's wort (*H. anagalloides*) and the erect species Scouler's St. John's wort (*H. scouleri*).

Ecology & Ethnobotany In general, Native Americans dried the whole plants and pulverized them into a meal, which was then used in cooking. The fresh leaves were also eaten. **Caution** The genus contains at least 6 species, worldwide which are poisonous. The toxin is a pigment, hypericin, which causes photosensitivity in the skin of animals who eat it. Upon exposure to light, such skin will form lesions, seep, itch, or fall off in more severe cases. There is no complete recovery from the after -effects of this poisoning.

DOGWOOD
Cornus sericea CORNACEAE

General Description Dogwoods are shrubs or semi-woody perennials with simple leaves that are opposite or whorled. A distinguishing characteristic of this family is the flower structure, which includes 4 to 5 sepals, petals, and stamens, all of which are attached at the top of the ovary. The flowers mature into red or white drupes.

Ecology & Ethnobotany The wood of this family is extremely hard and free of scratchy silica - so

much that jewelers reportedly used small splinters of it to clean out the pivot-holes in watches and opticians utilized it to remove dust from small-deep-seated lenses. Peeled twigs of any species can be used as tooth brushes.

Red-osier dogwood (*C. sericea*) is a shrub that occurs in many moist habitats below 8,000 feet. The species is highly variable with many local forms. The berries, which resist rot due to their low sugar content, provide long-lasting food for wildlife during winter. The most distinctive feature of this plant is its bright red bark. The Native Americans scraped, dried and smoked the inner lining of this bark, which is reported to have hallucinatory properties. The term osier refers to pliable twigs used in basket making.

STONECROP
Rhodiola integrifolia & Sedum   CRASSULACEAE

General Description Stonecrops are well-adapted to survival in shallow soil or on rocky outcroppings and look very similar to dudleyas. The succulent leaves and stems have a waxy coating to help reduce water loss. The reddish color of the foliage in some species is enhanced by sunlight and occurs most often in plants in hot exposed sites.

Ecology & Ethnobotany These are my favorite edible plants. The young leaves and stems of all species can be eaten as a salad or boiled as a potherb. I find them slightly tart and crisp - a wonderful addition to salads and trail snack. However, some species have emetic and cathartic properties, and can cause headaches. In an emergency, stonecrop can be eaten raw to allay hunger and thirst. The plants are best when collected before flowering since they tend to become bitter and fibrous in late summer. The green fleshy leaves are high in Vitamins A and C. The tubers can also be boiled and eaten.

Caution

Sedum has also been reported as being slightly astringent and mucilaginous. It is valuable in the treatment of wounds, ulcers, lung disorders, and diarrhea. As a field remedy for minor burns, insect bites, and other skin irritations, just squeeze the juice onto the affected area. Decoctions of the plant were also used for sore throats and colds, and as an eye wash.

JUNIPER
Juniperus   CUPRESSACEAE

General Description These are evergreen trees and shrubs, with opposite or whorled, scale-like or linear leaves. The male cones are small, and the female cones are larger and "berry-like."

Ecology & Ethnobotany As you will see, this is one of the important plants to know and junipers offer countless products. The berries and twigs can be used to make tea, season game, smoke fish, repel moths, soothe rheumatic pains, and kill infectious germs. The fleshy cones are edible raw, but taste better if dried, ground, and used as a flour, flour extender, or made into cakes. Cooking the flour with other foods can make it more palatable. The berries can also be roasted and ground, and used as a coffee substitute. The Swedes make an extract from the berries, which they generally eat with bread, in much the same way we use butter. In addition, the berries have been used to give gin its characteristic flavor. The boughs can be steeped in hot water for 5-10 minutes to make a beverage. Cooking them in an uncovered pot is recommended to allow the volatile oils to escape.

The shredded bark is an excellent tinder for primitive fire-starting techniques and can be used as bedding and padding. It is said that some Native American tribes ate the inner bark in times of famine. The inner bark was also used for clothes and mattresses and could be worked with the hands until soft enough to use for baby diapers or sanitary pads. Juniper oil extract can be rubbed on skin as an insect repellent, and to relieve pain in muscles and joints. The bark, roots,

twigs, and cones furnish red dyes. The white film covering the berries is a type of yeast that can be used to make a primitive sourdough starter.

SEDGE
Carex   CYPERACEAE

General Description Sedges are perennial, grass-like plants with creeping rhizomes, short rootstalks, or fibrous roots. The solid (not hollow) stems are usually triangular in cross section, and the leaves at the base of the stem are often reduced to scales. *Carex* is one of the largest genera in this area. Many of the species are difficult to tell apart, and a more technical key should be consulted. Mature fruit and a hand lens are necessary for positive identification of most species.

Ecology & Ethnobotany Any sedge should be tried for food. The young shoots or bases of the leaves are tasty additions to the diet

FLATSEDGE, CYPERUS
Cyperus   CYPERACEAE

General Description These are annual or perennial and grass-like. Leaves are mainly basal and three- ranked, and the stems triangular in cross-section. The spikelets are clustered in ball-shaped heads and the inflorescence consists of numerous heads borne on stalks radiating from the top of the stem and subtended by long, leaf-like bracts. The various species occur in wet, open soils along river banks, and the margins of lakes and ponds.

Ecology & Ethnobotany Several species bear tubers which are edible. They can be utilized in just about any way imaginable, from eating them raw to candying them or making a beverage from them. The inner base of the plant is edible raw.

TULE, BULRUSH
Schoenoplectus acutus & *Scirpus microcarpus*
CYPERACEAE

General Description Bulrushes are perennials with mostly solid, triangular or round stems. The leaves are grass-like or reduced to sheathing scales. The flowers consist of an ovary, 3 stamens, and 2 to 6 bristles at the base of the ovary. Each flower is subtended by a firm, thin scale, and they are spirally arranged in compact, egg-shaped or elliptical spikelets. The inflorescence is subtended by 1 to several bracts that are small or elongated and leaf-like, and spreading or erect and appearing to be part of the stem. The fruit is an achene. The species are generally found in marshes, wet meadows, along streambanks and pond margins.

Ecology & Ethnobotany In general, large rhizomes of *Scirpus* species can be eaten raw or crushed into a sweet tasting flour and used as mush. The seeds are also edible, best when used as flour after grinding. The pollen may be gathered and pressed into cakes and baked.

Bulrush stems, roots, and leaves can be used as the foundation and twining material in twined baskets. They were used extensively by Native Americans for making cordage, sandals, baskets, and mats. Waterproofed "water bottles" can be made from the baskets by coating the inside with asphaltum or pine sap. The stems were used for sleeping mats, padding, thatching dwellings, skirts, and sandals. Duck-shaped decoys were also made when hunting.

 WESTERN BRACKENFERN
Pteridium aquilinum   DENNSTAEDTIACEAE

General Description This is a medium- to large-sized plant with decompound, broadly triangular leaves up to 7 feet long including the stipe. The stipes are green or yellowish and there are fine white hairs. The sori are marginal and continuous, and partially covered by the recurved leaf margins. This is a widely distributed species found in open woods, rock slides or slopes in damp or dry places, up into the high mountains.

Ecology & Ethnobotany The young fronds, or fiddleheads of brackenfern can be collected, boiled and dried in the sun. The dried product can then be used as a food. Old fronds may be poisonous in large. The starchy rhizome (underground horizontal stem) is edible after roasting or boiling but is usually tough. The leaves can be used as one of the protective plant layers for pit cooking. **Caution** While this plant has traditionally been accepted and harvested as a suitable edible, there is evidence indicating that eating sufficient quantities over a period of time may be dangerous to your health.

RUSSET BUFFALOBERRY
Shepherdia canadensis   ELAEAGNACEAE

General Description This is a low spreading shrub with opposite leaves, each with a dark green upper surface and lighter underside that is covered with tiny, brown scales. The inconspicuous flowers are yellow-green; the fruits range in color from yellow to bright red. Buffaloberry is common in forested environments,

from the montane to subalpine/lodgepole pine forests. It is also found in recently burned areas.

Ecology & Ethnobotany Another common name for this species is "soapberry." The berries contain a significant amount of saponin which not only gives the plant its bitter taste, but also whips up into a frothy mass called "Indian Ice Cream."

Native Americans used the berries of these plants extensively, both fresh and dried. The berries of this species are at first pleasant, then the soap-like bitterness prevails. We enjoy cooking them with sweeter tasting berries such as thimbleberries (*Rubus*) and serviceberries (*Amelanchier*), and a large amount of sugar. We found the berries to be somewhat unattractive for general use, but a valuable consideration in emergencies. The berries taste better after a few good frosts during the fall. They can also be used in the making of pemmican or jelly. Dried into cakes, the berries can be stored for winter.

HORSETAIL
Equisetum   EQUISETACEAE

General Description In general, these are rhizomatous ferns with hollow, grooved, regularly jointed stems that are impregnated with silica. The leaves are reduced in size, appearing as a series of teeth around a joint. Spores are produced in cone-like structures atop the stems. They are found in moist soil along streams and rivers, marshes, and other damp habitats.

Ecology & Ethnobotany
Although all species are useful and identical in application, common horsetail is the most popular. The tough outer tissue can be peeled away and sweet inner pulp of all species can be eaten in small amounts. In large quantities, defined as greater

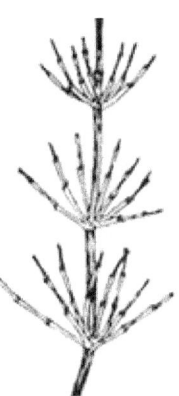

than 20% of body weight by some authorities, they can be toxic. Certain chemicals in this plant are said to destroy specific B Vitamins such as thiamine. The enzyme thiaminase is apparently responsible for the poisoning. Cooking destroys this enzyme and renders the plants safe for consumption. The tuberous growth on the roots (actually rhizomes) can be eaten raw in the early spring or boiled later in the season.

In the fall, the stems become impregnated with silicon dioxide and can be used to scour pots and pans or as a type of sandpaper for wood.

KINNIKINNICK
Arctostaphylos uva-ursi   ERICACEAE

General Description *Arctostaphylos* include the manzanitas of the Pacific Coast. This is a low, mat-forming shrub with reddish to brown stems that root at the nodes. The spoon- to lance-shaped leaves are leathery. The flowers are urn-shaped and the bright red berries often persist through the winter. It is found in

open areas with dry to well-drained soils from low to high elevations. The genus name means "bear grape" and refers to the fondness shown by bears for the fruits of these shrubs, many of which are known as Bearberry.

Ecology & Ethnobotany The berries of all manzanita species are edible. They may be eaten raw, and it is suggested that they not be eaten in large quantities since they may be hard to digest. Constipation or indigestion are common maladies of eating too much. The berries can also be stewed, or dried and ground into meal and cooked as mush. A cider can also be made from the berries. The seeds alone can be collected and ground into meal too.

SPICYWINTERGREEN
Gaultheria   ERICACEAE

General Description Two species of *Gaultheria* may occur in the Sawtooth Country and are dwarf evergreen shrubs that form small mats with leaves that are broadly egg-shaped to elliptical. The bell-shaped flowers are white to pink and the berries are red. These plants can be found in moist areas at mid elevations.

Ecology & Ethnobotany The small, red fruits of *G. humifusa* (alpine spicywintergreen) are edible raw or cooked, and can be made into jams, wines, or pies. The young tender leaves are suitable as greens and have a wintergreen flavor.

In Native American medicine, the plants were used for treating aches and pains and to help with breathing while hunting or carrying heavy loads. The leaves of these species yield oil upon steam distillation. This "oil of wintergreen" (methyl salicylate) is a folk remedy for body aches and pains, and is known for its astringent, diuretic, and stimulant properties. **Warning** The wintergreen flavor in the plants is due to the presence of oil of wintergreen, which, if taken in excess can be toxic, especially to children. In small amounts, such as in wintergreen tea, there is little danger. Children who are allergic to aspirin (a related drug) should not eat the plant or berries, or even handle the plant.

☠ MOUNTAIN LAUREL
Kalmia polifolia   ERICACEAE

General Description This is a branched, evergreen shrub that spreads by short rhizomes and layering. The leaves are opposite, lance-shaped to elliptical and have in-rolled margins. The flowers are rose-colored and bowl-shaped. The species occurs in the mid- to upper elevations, usually in moist to wet, acidic soils, often along creeks.

Ecology & Ethnobotany The *toxicity of Kalmia is legendary.* Some Native Americans used it as a suicide plant. Game birds and livestock may be poisonous to eat if they have ingested the leaves. According to Peter Kalm (1715-1779), after whom the genus is named;

"...sheep are especially susceptible, while deer are unharmed. Though the flesh of affected animals is

apparently not contaminated, the intestines will cause poisoning if fed to dogs so that they become quite stupid and as it were intoxicated and often fall so sick that they seem to be at the point of death."

All Kalmia species should be considered poisonous. They contain andromedotoxin, which causes a slow pulse, low blood pressure, lack of coordination, convulsions, progressive paralysis, and death. The honey made by bees from these plants is also poisonous.

HUCKLEBERRY, BLUEBERRY
Vaccinium   ERICACEAE

General Description These are small to mid-sized shrubs with deciduous leaves. The twigs are often angled. The small flowers are urn-shaped, and fruits many seeded berries. They can be found on well-drained sites, from wet meadows and around lakes up to the timberline.

Ecology & Ethnobotany *Vaccinium* berries can be eaten raw or be dried in the form of cakes for future use. The various species range in taste from sweet to tart. Hybridization between the species is known to occur, but the fruits are still edible. The berries have also been used as fish bait since they look very similar to salmon eggs. The leaves can be dried to make a tea. The leaves and berries are high in Vitamin C.

☠ TURKEY-MULLEIN
Croton setiger   EUPHORBIACEAE

General Description Formerly known as *Eremocarpus setigerus*. turkey-mullein is a low spreading gray plant with hispid stems and pubescent leaves. Leaves are thick, ovate in shape and palmately veined. This is an abundant weed encountered in disturbed areas in the Sawtooth Country at the lower elevations.

Ecology & Ethnobotany The herbage of the plant is regarded as poisonous. The poison, presumably diterpenes, was used to stupefy fish in the past.

LICORICE
Glycyrrhiza lepidota   FABACEAE

General Description This plant has deep, extensive rhizomes which give rise to erect, rigid stems growing up to 48 inches tall and that are covered with stalked or sessile glands. Yellow-brown glands also dot the underside of the 11-19 lance-shaped leaflets. The pale yellow to greenish-white flowers are crowded along the terminal potion of the long, naked stalk arising from the upper leaf axils. The short pods are brownish and have dense bristles. Usually found in moist, sandy soils, river banks at the lower elevations.

Ecology & Ethnobotany The plant contains glycyrrhizin, sugar and other chemicals used in medicine as a mild laxative, a demulcent, and a flavoring to mask the taste of other drugs. It is also used in confections, root beer, and chewing tobacco. Licorice has been used in the treatment of asthma, stomach ulcers, bronchitis, and urinary tract disorders. The plants were chewed by

Native Americans and used as a flavoring. Warning - continual use of this plant in large doses may cause water retention and elevated blood pressure.

LUPINE
Lupinus　　FABACEAE

General Description There are many species of lupine. In short, they are showy perennial or annual herbs with palmately compound leaves. Flowers range in color from blue, violet, rarely white, to rose and occur in elongated narrow inflorescences. The pods are flattened and usually hairy. They are found on open slopes and meadows up into the alpine zone.

Ecology & Ethnobotany Due to their abundance in some areas, it is *important to quickly dismiss these plants as being edible.* In many field guides, the pea-like seeds have been wrongly recommended by the authors as a substitute for peas. Lupines possess many complex alkaloids and should be considered poisonous.

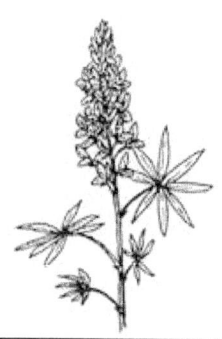

SWEETCLOVER
Melilotus   **FABACEAE**

General Description Sweetclover are strongly tap-rooted perennial or annual herbs. The leaves are divided into three fine-toothed wedge-shaped leaflets. The white or yellow flowers are loosely arranged in an inflorescence and the pods are thickly spindle-shaped. The species are usually found in disturbed habitats at the lower elevations.

Ecology & Ethnobotany The young leaves (before the flowers appear) may be eaten raw or boiled. The fruit may be used as seasoning for soups. The older leaves are toxic and should be avoided. Improperly dried sweetclover will easily mold and should be considered poisonous. Molding sweetclover mixed in hay has killed many cattle.

CLOVER
Trifolium   **FABACEAE**

General Description There are many species of clover. In general, they are annual and perennial plants from rhizomes with leaves that are divided into three or more leaflets. The flower colors range from white, pink, yellow, red, or purple, and the seed pods are round to elongated. They are found in various habitats at all elevations.

Ecology & Ethnobotany All species are nutritious and high in protein, but the flower heads and tender young leaves are hard to digest raw and may cause bloating. To improve digestibility of the plants, soak them in salt water for several hours or overnight.

Leaves prepared this way may be dried and stored for future use. The dried flower heads and seeds can be ground into a flour substitute or extender.

VETCH
Vicia   **FABACEAE**

General Description Vetch are annual or perennial herbs with trailing to climbing stems. Leaves are pinnately divided with tendrils in place of terminal leaflets. Found in waste places at lower elevations. *Vicia* closely resembles *Lathyrus* and requires close examination of the stipules. The stipules of *Vicia* are usually cut into narrow lobes, whereas the stipules of *Lathyrus* are entire to dentate. *Both genera should be avoided and considered poisonous despite some people suggesting otherwise.*

Ecology & Ethnobotany Many species contain toxic compounds and therefore *should be considered poisonous*. However, several authorities state that the young stems and seeds can be boiled or baked. The seeds of some species contain compounds producing toxic levels of cyanide when digested.

ELKWEED, MONUMENT PLANT
Frasera speciosa   GENTIANACEAE

General Description This plant can grow up to 6 feet tall. The flowers are greenish yellow, spotted with purplish, and occurring in elongated, terminal group. In the Sawtooth Country, this plant can be found on dry slopes above 6,800 feet.

Ecology & Ethnobotany The fleshy root of this species can be eaten raw, roasted, or boiled. But, because the root is very bitter, we suggest mixing it with salad greens.

RED-STEMMED STORKS-BILL
Erodium cicutarium   GERANIACEAE

General Description Red-stemmed storks-bill is a low growing annual with mostly basal, finely dissected, fernlike, pinnately divided leaves. The flowers are small, pink and mature into the distinctive "stork's bill" fruit. This is an introduced plant that is widespread on disturbed sites at low to middle elevations.

Ecology & Ethnobotany The leaves of this species can be eaten raw in salads or cooked as a potherb. They are particularly palatable when picked young and have a parsley-like taste. We find it nicely compliments an otherwise bland wild salad and provides a good source of Vitamin K. It is uncertain whether other species of

Erodium are edible and it is not recommended.

GERANIUM
Geranium   GERANIACEAE

General Description These are annual or perennial herbs that are hairy. The leaves are mostly basal, and the flowers are showy, with 5 petals and sepals, and 10 stamens. The mature fruits are spirally coiled. The species in the Sawtooth Country can be found in wet meadows or dry, open forests. Species include: Richardson's geranium (*G. richardsonii*) and sticky purple geranium (*G. viscosissimum*)

1. Petals white with pink or purple veins; inflorescence glandular usually with purple-tipped hairs ----- ***G. richardsonii***
1. Petals pink or purple; inflorescence not glandular, or glandular with yellow or whitish-tipped hairs ----- ***G. viscosissimum***

Ecology & Ethnobotany The leaves and flowers of most species can be eaten, but because of their astringent properties and texture, they are not a choice edible. We find that they are best when tossed in with

other greens in salads or steamed as potherbs. In any case, the leaves are better treated as a filler to stretch supplies of tastier and less abundant greens. The leaves can also be chopped and added to soups, thereby blending flavors making the leaves more acceptable. Leaves do toughen with age but are still palatable in stews. Geranium leaves are similar looking to monkshood (*Aconitum*), so positive identification of the flowerless plants is important.

CURRANT, GOOSEBERRY
Ribes   GROSSULARIACEAE

General Description Members of this genus are shrubs. The species that have prickles on the stems and bristles on the fruit are commonly called gooseberries. Those without prickles on the stem or bristles on the fruit are currants. Leaves are palmately veined and shallowly or deeply lobed. The five petals are smaller than the sepals and usually narrowed to a claw-like base. Fruit is a berry.

Ecology & Ethnobotany The berries of all species are edible raw, and none are known to be poisonous. However, we have come across some unpalatable species, berries with an unpleasant odor and a taste to match. The berries are high in Vitamin C and one of the richest plant sources for copper. One method of collecting them in bulk, is by shaking the bushes over sheets of plastic or blankets. Those that are too sour or spiny become more palatable if they cooked or dried. In regard to the fruits with bristles, one can also roll the berries on hot coals in a basket until the bristles have been singed off. When dried, the berries are a great trail snack. The dried berries can also be mixed with meat to make pemmican. The berries contain enough natural pectin to make jelly. The seeds also contain large quantities of gamma-linolenic acid and many herbalists use this oil to treat skin conditions, asthma, arthritis, and premenstrual syndrome. The nectar-filled flowers are considered good trail snacks. The wood makes good arrow shafts.

MARE'S-TAIL
Hippuris vulgaris   HIPPURIDACEAE

General Description Mare's-tail is a common plant in the main mountain chains of the western United States. The plant at first glance resembles an immature horsetail (*Equisetum*), but they are unrelated. Horsetails reproduce by spores and have stems that can be quickly pulled apart. The flowers of mare's-tail are small and inconspicuous. The plant is found in the margins of shallow waters from ponds to lakes to streams. It can also be found in marshy and swampy areas, roadsides, and irrigation ditches.

Ecology & Ethnobotany The whole plant is edible when prepared as a potherb. The plant parts are tender and can be gathered in any stage, even in winter.

Ancient herbalists are said to have employed mare's-tail for internal and external bleeding.

WILD MOCKORANGE
Philadelphus lewisii   HYDRANGEACEAE

General Description This is the State Flower of Idaho and is also known by the common name of Syringa. This is a loosely branched shrub or tree with white flowers. It grows on rocky slopes and canyons below 6,000 feet. The genus is named for the Egyptian King Ptolemy Philadelphus.

Ecology & Ethnobotany The wood of mockorange is strong and hard and does not crack or warp. It is an excellent wood for making bows and arrows. The leaves and flowers foam into lather when bruised and rubbed with hands and can be used for cleaning the skin. The plant is otherwise considered poisonous.

WATERLEAF
Hydrophyllum   HYDROPHYLLACEAE

General Description Two species of waterleaf occur in the Sawtooth Country - ballhead waterleaf (*H. capitatum*) and Fendler's waterleaf (*H. fendleri*). They are somewhat fleshy perennial herbs with leaves that are pinnately divided. Flowers are white to bluish. They are found in moist soils from the foothills to alpine environment.

1. Inflorescence dense, usually globose; peduncles longer than petioles ----- *H. capitatum*
1. Flowers not in globose heads, inflorescence more or less open; peduncles shorter than petioles ----- ***H. fendleri***

Ecology & Ethnobotany The young shoots, leaves, and flowers of both species can be eaten raw, or these and the roots may be cooked and eaten. We find them exceptionally good in salads, or when eaten as a trail nibble. They do have a texture that takes some getting used to.

The leaves can be used as a protective dressing for minor wounds and are slightly astringent. As a poultice, it can be used for insect bites and other minor skin irritations.

BRANCHING PHACELIA
Phacelia ramosissima   HYDROPHYLLACEAE

General Description There are about 11 species of *Phacelia* in the Sawtooth Country, but it is branching phacelia that has used as food in the past. The genus is challenging to work with in the field and the species sometimes hybridize.

Branching phacelia is a coarse perennial that grows up to 40 inches tall and is covered with long, bristly hairs. The leaves are oblong to ovate in shape and pinnate into toothed divisions. The flowers occur in a densely coiled inflorescence. Corolla is bell-shaped and bluish in color. This phacelia is common in shaded canyons and forest below 8,000 feet and flowers from May to August.

Ecology & Ethnobotany This phacelia species is edible and can be cooked as greens.

ROCKY MOUNTAIN IRIS
Iris missouriensis   IRIDACEAE

General Description The stems of this plant are about two feet tall with several grass-like leaves. The flowers are showy with three drooping blue sepals and three petals that are slightly smaller than the sepals. Rocky Mountain iris can be found in meadows, wet or moist areas at low to mid elevations.

Ecology & Ethnobotany Members of this genus contain irisin, an acrid

resin concentrated mainly in the rhizomes, and present in the foliage and flowers. People who raise irises sometimes develop a skin rash from handling the rhizomes. The rootstock produces a burning sensation when chewed. If eaten in quantity, irises will cause diarrhea and vomiting. The poisonous rootstalks were used by Native Americans in a mixture of bile to poison arrow points. The leaves can be used to make crude cordage in making nets and snares.

HORSE MINT
Agastache urticifolia　　**LAMIACEAE**

General Description This is a tall perennial growing 3 to 6 feet tall. The leaves are opposite, ovate in shape, 1 to 3 inches long and 1½ inches wide. They are also coarsely toothed on the margins. Flowers occur in dense whorls and form a terminal spike, 1½ to 6 inches long. The calyx is green or rose, and the corolla is 2 lipped, rose or violet, and 3/8 to 5/8-inch long. In the Sawtooth Country, horse mint grows in moist places below 10,000 feet. Flowers from June to August.

Ecology & Ethnobotany The seeds of horse mint may be eaten raw or cooked, and the leaves can be used in tea or for flavoring stews. Some Native Americans in California drank an infusion made from the leaves to relieve rheumatic pain, and for indigestion and stomach pains. The plant is said to have mild sedative qualities. The mashed leaves were made into a poultice and applied to swellings.

HENBIT
Lamium amplexicaule　　LAMIACEAE

General Description This is a weedy annual from Eurasia. The leaves are rounded to heart-shaped with shallow, rounded teeth on the margins. The flowers are pinkish-purple arising from the leaf-like bracts. It is found in waste places and fields at the lower elevations.

Ecology & Ethnobotany The entire plant is edible. We found it best when added to other salad plants. The plant is mildly astringent and a tea from the leaf and flower is used by herbalists for minor internal or external bleeding, to relieve diarrhea, and other digestive problems. *Caution* - In most cases the plant may be found in areas that are often sprayed with herbicides and pesticides.

BUGLEWEED
Lycopus   **LAMIACEAE**

Lycopus

General Description The two species in the area are perennial herbs with rhizomes. Of the three species, *L. americanus* (American waterhorehound) is the more common. The small flowers are pinkish-purple, whorled in the axils of the upper leaves. They occur in wet to moist areas of lakes and riverbanks. Unlike other mints, bugleweeds do not have the mint-like odor.

Ecology & Ethnobotany The leaves of American waterhorehound are edible raw but are usually tough and bitter.

WHITE HOREHOUND
Marrubium vulgare   **LAMIACEAE**

General Description This is a woolly perennial herb with bitter sap. The leaves are wrinkled and toothed. The flowers are small, white and occur in dense whorls. Horehound is a common weed of waste places and fields at the lower elevations.

Ecology & Ethnobotany The most famous use of this plant is horehound candy and is used to soothe sore throats and coughs. A tea from the dried leaves and flowers is also used, but because of the extreme bitterness of the herb, it is obvious why it tastes better in the form of a candy.

WILD MINT, SPEARMINT
Mentha   LAMIACEAE

General Description These are distinctly aromatic perennial herbs with rhizomes. The flowers are arranged in whorls. The species are generally found in moist habitats.

Ecology & Ethnobotany The fresh or dried leaves of *M. arvensis* (wild mint), *M. canadensis* (Canadian mint), and *M. spicata* (spearmint) can be steeped in hot water for a tea. They have also been used as flavoring agents for soups, meat, and pemmican. The young leaves can also be added to salads and soups. The plants are high in Vitamin A, C, K, and minerals iron, calcium and manganese. It is an appetite stimulant and digestive aid.

CATNIP
Nepeta cataria   **LAMIACEAE**

General Description This is a tap-rooted perennial that feels like felt to the touch. Leaves are triangular-shaped and coarsely toothed. The flowers are blue or yellowish-white in terminal, spike-like inflorescences. This introduced species from Europe is now widespread across North America and occurs in waste and disturbed places, and along irrigation canals at the lower elevations.

Ecology & Ethnobotany The nutritious young leaves and buds can be added to salads. The dried leaves

and flowers make an excellent tea and are high in trace minerals and vitamins. The tea is a subtle, relaxing sedative on humans. As with other mint teas, it is also a carminative, and is soothing to an upset stomach.

SELF-HEAL
Prunella vulgaris **LAMIACEAE**

General Description This perennial grows 4 to 20 inches tall, and has opposite, lanceolate ovate shaped leaves that are 1to 2 inches long. The herbage is glabrous to short pubescent. Flowers occur in a dense, terminal spike in the axils of round, membranous, purple tinged bracts. The calyx is purplish and the corolla is 2 lipped, violet, and 3/8 to ¾ inch long. Self-heal is common in moist woods and flowers from May to September.

Ecology & Ethnobotany The entire plant is edible, raw or cooked. However, we found that it is the young and tender plants collected in the early spring that are best. The crushed leaves can be used fresh or dried to make a tea.

DUCKWEED
Lemna **LEMNACEAE**

General Description These are small plants, often not much larger than a pinhead. and float on slow or stagnant waters. Unlike the ordinary leaves of most plants, each duckweed frond contains buds from which more fronds may grow. They have small thread-like root hairs that obtain nutrients from the water.

Ecology & Ethnobotany Under survival conditions, duckweed can provide copious and palatable material for salads. *Caution is advised*: the waters in which these plants are found may be contaminated.

WILD ONION
Allium   LILIACEAE

General Description The many species of onions arise from bulbs, and all have the characteristic distinct onion odor. The small flowers are clustered together in umbels. In the Sawtooth Country, at least 14 species of onions are found in a variety of habitats from the low elevations to the high alpine zone.

Ecology & Ethnobotany All species are edible. The bulbs may be eaten raw, boiled, steamed, creamed, in soup, and are especially good when used as a seasoning. Ingestion of large amounts of onions, including the cultivated ones, can cause poisoning or cause goiter, but are otherwise not known to be harmful. The seeds and leaves can also be eaten. Onions will keep a long time, because the skin dries and preserve the flesh inside. Onions contain large amounts of some important micronutrients, more Vitamin C than an equal weight of oranges, and more than twice as much Vitamin A as an equal weight of spinach.

Medicinally, people have taken advantage of onions natural antiseptic properties by applying the juice to wounds to prevent infection. The juice was also used as an insect repellent when rubbed over the body. The onion smell apparently has some beneficial effects on the circulatory, digestive, and respiratory systems.

MARIPOSA LILY
Calochortus   LILIACEAE

General Description Six species of *Calochortus* occur in the Sawtooth Country. They are characterized as perennials from bulbs, with tulip-like flowers that are few and showy. These species can be found in dry open places from low to mid elevations.

Ecology & Ethnobotany Any of the species in the area are edible. There seems to be practically no way in which these plants cannot be prepared for food. The whole plant can be used raw, cooked, or whatever. The biggest threat to these plants are overharvested by people. Consider these plants only in an emergency.

CAMAS
Camassia quamash   LILIACEAE

General Description This species was considered an important food plant and was extensively utilized by Native Americans. Arising from a bulb, it has bright blue to violet, 6-parted flowers in a showy spike-like raceme. The leaves are basal and grass-like. The plant can be found in meadows, marshes, grassy slopes, and fields from low to mid elevations.

Ecology & Ethnobotany Camas bulbs can be dug out of the ground from late July through September with digging sticks. The black outer covering of the bulb is removed, and the white bulbs are then steamed or cooked in pits for 24 hours or more. When cooked this way, the bulbs turn dark brown, become quite moist and soft, and sweet.

Cooking is required because the plant contains a carbohydrate called inulin. Inulin is not very digestible or very palatable in its "raw" form. Cooking is necessary to chemically breakdown the inulin into its component fructose sugar. Common in fruits and honey, fructose is both easily digested and sweet tasting. After cooking, the camas bulbs can be mashed and dried into cakes for storage. They are considered to be more nutritious than potatoes. The bulbs can also be boiled down into syrup.

Too much camas, however, is both an emetic and purgative. Caution When collecting bulbs, be aware that the poisonous death camas (*Toxicoscordion*) may be in the area too.

GLACIER LILY
Erythronium grandiflorum   LILIACEAE

General Description. This is the only species in the Sawtooth Country. The bright yellow flowers hang from a slender stalk. The two flat leaves are sheathing at the base. Glacier lily can be found up to and above timberline, blooming at the edge of melting snowbanks.

Ecology & Ethnobotany These beautiful flowers are seldom abundant, so glacier lily *should only be considered in extreme emergencies*. Harvesting destroys the plant. The young plants can be boiled as a potherb. The seed pods can be eaten raw or cooked. Eating the seed pod will not destroy the plant as long as you spread the seeds around first. Corms may be eaten raw but are better if boiled for at least 20 minutes. They can also be dried and stored for future use. The corms contain inulin, which is inedible raw. Normally they would be pit cooked for an extended period of time.

FRITILLARY, YELLOW BELLS
Fritillaria   LILIACEAE

General Description Two species occur in the Sawtooth Country - spotted missionbells (*F. atropurpurea*) and yellow bells (*F. pudica*). In general, they are glabrous, perennial herbs from bulbs with numerous white bulblets around the base. The flowers are usually nodding with similar petals and sepals (tepals). Fritillary can be found in open areas, forest, meadow, and grassland habitats at low and middle elevations.

1. Flowers yellow, fading to red ----- ***F. pudica***
1. Flowers brown or purple, mottled with yellow ----- ***F. atropurpurea***

Ecology & Ethnobotany The bulbs of this genus have been a staple for Native Americans since prehistoric times. Bulbs of all species are edible raw or cooked but are relatively rare and should be considered only in an

emergency.

FALSE SOLOMON'S-SEAL
Maianthemum   LILIACEAE

General Description These are annual herbs with extensive, horizontal rootstalks. Leaves are alternate and sessile or on short petioles. Two species occur in the Sawtooth Country - feathery false solomon's-seal (*M. racemosum*) and starry false solomon's-seal (*M. stellatum*). Both species are generally found above 4,500 feet in moist or shady areas. They may be identified using the following key.

1. Perianth segments, less than 1/8-inch long; fruit is a red berry ----- ***M. racemosum***
1. Perianth segments ¼ inch long; fruit a purple to black-colored berry ----- ***M. stellatum***

Ecology & Ethnobotany Both species have edible berries that are not especially palatable. If eaten in quantity they can act as a laxative. Cooking the berries removes much of the purgative elements making them a bit more palatable. They are high in Vitamin C. The young shoots and leaves can be used like asparagus or eaten as a potherb.

False solomon's-seal have starchy rootstocks that may be eaten.

However, the rootstocks must be soaked overnight in lye. Native Americans in Canada, used the white ash from their fire pits instead of lye, which supposedly removed the bitterness. The roots are then boiled and rinsed several times to remove the lye.

The mashed rootstock of starry false solomon's-seal was thrown into a stream as a fish stupifier, making the fish easier to catch.

☠ CLASPLEAF TWISTED-STALK
Streptopus amplexifolius   **LILIACEAE**

General Description Twisted stalk is a perennial herb with creeping rootstocks. The sessile or clasping leaves are alternate, elliptical to ovate in shape, and the flowers are yellowish-green. The 1-2 pendant flowers hang from the axils of the upper leaves on stalks that are bent in the middle. It is common in moist soil and along streams and thickets in the montane and lower subalpine zone, they are often associated with genera such as *Maianthemum* and *Actaea* (baneberry).

Ecology & Ethnobotany In terms of edibility, these plants have escaped mention in many guides but are indeed safe. The new spring shoots and clasping young leaves can be eaten raw or added to salads and taste somewhat like cucumbers. The berries, often referred to as watermelon berries are somewhat laxative if eaten in excess but may be eaten raw or cooked in soups and stews. They are sometimes referred to as "scooter berries," because if you eat too much you can find yourself "scooting" off to the bathroom. The species are easy to grow in wild gardens. The stems were used in poultices for cuts.

Warning Anyone wishing to use the young shoots of twisted stalk should be very careful to identify it correctly. At the shoot stage, these plants resemble the highly toxic corn lily (*Veratrum*).

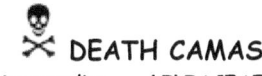 DEATH CAMAS
Toxicoscordion   LILIACEAE

General Description Previously included in the genus *Zigadenus*. Three species occur in the Sawtooth Country. They are glabrous perennials with bulbs and grass-like leaves. The cream-colored to greenish white flowers are stalked and subtended by narrow bracts in an elongated inflorescence.

Ecology & Ethnobotany Death camas is very poisonous plant if ingested. The alkaloids, primarily concentrated in the bulbs, can cause muscular weakness, slow heartbeat, subnormal temperature, stomach upset with pain, vomiting, and diarrhea, and excessive watering of the mouth. Death camas should not be confused with the edible camas (*Camassia*), which does occur in our area, but which formed a staple food for aboriginal peoples in the Pacific Northwest. It is also difficult to distinguish death camas from other edible plants, including wild onion (*Allium*), Mariposa lilies (*Calochortus*), fritillaries (*Fritillaria*), and brodiaeas (*Brodiaea*) prior to flowering.

PACIFIC TRILLIUM
Trillium ovatum   LILIACEAE

General Description This is a low, glabrous herb from short, fleshy rootstocks. The stem bears a whorl of three sessile, broadly egg-shaped leaves, and the single stalked flower is borne above them. The sepals are green, and narrower and shorter than the snow-white petals. *Trillium* is common in moist, rich soils of forests from the foothills to the subalpine zone.

Ecology & Ethnobotany The stem and leaves may be boiled and eaten as greens. The leaves should be collected before they fully unfold because when the flowers appear, the leaves become bitter. The berries and roots are inedible, possessing emetic properties. Because of their relative rarity and beauty, refrain from using them except in an emergency situation.

LARGEFLOWER TRITELEIA
Triteleia grandiflora   LILIACEAE

General Description The flowering stems are erect and the leaves are few, basal, and grass-like. The leaves often wither away before the flowers appear. The flowers occur in an umbel and the segments all look alike. The fruit is a capsule. These plants occur on slopes and flats below 5,000 feet.

Ecology & Ethnobotany The corms of most *Triteleia* species are edible raw but are somewhat mucilaginous. It is better if they are boiled for a few minutes or roasted. They can

also be mashed and dried for future use in stews. Since the corms grow deep, it is usually easiest to harvest them with digging sticks.

The crushed corms were used as a paste which was smoothed over the sinew backing on bows. The paste was also used to bind paint pigments to hunting bows. *Triteleia* corms and flowers were used as soap and shampoo.

☠ CORN LILY, FALSE HELLEBORE
Veratrum   LILIACEAE

General Description These are characterized as large, coarse, leafy-stemmed herbs with thick rootstocks. The leaves are broad, clasping, and strongly veined. The numerous green to dull white flowers are arranged in a large panicled inflorescence. False hellebore is found in wet meadows and forest openings from the middle elevations into the alpine zone. In the Sawtooth Country, corn lily is common in wet meadows and along stream banks, particularly at higher elevations; usually found below 11,000 feet.

Ecology & Ethnobotany These plants *are very poisonous if ingested* and have an inconsistent mixture of several powerful alkaloids. Some of the symptoms include depressed heart action, salivation, headache, burning sensation in the mouth, slowing of respiration, and death from asphyxia. These violent symptoms of poisoning may occur within 10 minutes. Avoid any use of the plant that involves ingestion. In some cases, just handling Veratrum can cause severe itchiness and irritation. Even nectar in the flowers is poisonous to insects and can cause serious losses among honeybees. **Personal observation** - I have observed many scout leaders mistakenly recommend their scouts use the leaves to wrap food for pit cooking. *DO NOT FOLLOW THIS FOOLISH ADVICE!*

BEARGRASS
Xerophyllum tenax   **LILIACEAE**

General Description This is a stout perennial up to five feet tall. The stems arise from large clumps of bluish-green, wiry, saw-edged, grasslike leaves that are up to two feet long. The cream-colored flowers are borne in a hemispheric terminal inflorescence. Common beargrass is found in mid to lower alpine open slopes and forests. The thick, shallow rhizome can remain vegetative for many years, then flowers and dies. The author has encountered this species on the northeast side of the Sawtooth Mountains and northwards.

Ecology & Ethnobotany The fibrous roots of common beargrass can be eaten after roasting or boiling. Although the sharp leaves are not very pleasant to handle, when dried and bleached they can be used for weaving baskets and clothing. The baskets are particularly pliable and durable. Lather from the roots was used to bathe sores.

FALSE MERMAIDWEED
Floerkea proserpinacoides   **LIMNANTHACEAE**

General Description False mermaidweed is a slender annual with succulent stems. The alternate leaves are pinnately divided into 3-5 oblong leaflets. The white flowers are stalked and borne singly in the leaf axils. Each bisexual flower has 3 sepals twice as long as the white petals. There are 3 or 6 stamens and the fruit consists of 2 rounded, bumpy lobes joined at the base. This species is found in moist, shaded habitats, especially under shrubs such as sagebrush upwards into lodgepole pine forest.

Ecology & Ethnobotany The root system consists of a slender branching taproot. This plant spreads by re-seeding itself, and it often form colonies at favorable sites. The small inconspicuous flowers can attract flower flies and small bees. The foliage is not known to be toxic and is edible to mammalian herbivores. We have sampled the stems and leaves of this plant and found them to be rather spicy and they were an okay addition to our wild salad.

LEWIS' FLAX
Linum lewisii **LINACEAE**

General Description This much branched annual has blue or rarely white flowers and alternate, sessile, and linear leaves. In the Sawtooth Country this species grows on open slopes from 1,000 to 9,500 feet. Flowers from May to July.

Ecology & Ethnobotany The seeds contain a cyanide compound but are edible after roasting them. They have a high oil content that contains essential fatty acids that are very much needed in our daily lives, plus they add an agreeable flavor to cooked foods. The stems are a source of linen, a fabric used for clothing.

BLAZINGSTAR
Mentzelia **LOASACEAE**

General Description The common name, blazingstar, actually describes only the tall and showy species. The little-known annuals are delicate and inconspicuous. The herbage of these plants is generally rough to the touch or with barbed, sometimes stinging hairs. The alternate, simple, entire to pinnately cleft and brittle leaves adhere to any foreign object contacted. Hence the other common name of stickseed. Flowers are bisexual, radially symmetrical, and borne singly or in a convex or flat-topped cluster.

Ecology & Ethnobotany *Mentzelia* was considered an important food source in many places of the West. The seeds of most species may be parched, and then ground into flour. The seeds can be stored for future use. The Hopi Indians in the southwest parched and ground the small, oily seeds of *M. albicaulis* into a fine, sweet meal and ate it in pinches.

COMMON MALLOW
Malva neglecta   MALVACEAE

General Description The species is distinguished by its distinctive fruit and seeds, rather than their leaves and flowers. It is an introduced annual herb that is usually found in waste places at the lower elevations.

Ecology & Ethnobotany The entire plant of *M. neglecta* is edible. The young leaves are particularly good in salads or cooked up as a potherb. The plant is, however, very mucilaginous, and it is often used to thicken soup and may take a little getting used to. Eaten in large amounts, however, may cause digestive disorder. The immature fruits (which look like cheese) can also be eaten raw or added to soups.

WILD BUCKBEAN
Menyanthes trifoliata   MENYANTHACEAE

General Description This is a perennial marsh herb with creeping root-stocks. The leaves have long petioles and are all basal and divided into 3 leaflets. The whitish flowers are small, star-shaped, and crowded into a short inflorescence. The species can be found in bogs and lakes. The species was at one time placed in the Gentianaceae (Gentian family).

Ecology & Ethnobotany The herbage and rhizome of the plant are bitter, but we found that the rhizome can be made palatable when collected in early season and boiled in several changes of water. A nutritious flour can also be made from the rhizome by drying, crushing, and leaching it thoroughly. The fresh plant eaten raw may cause vomiting. The fruit has no known use.

☠ WOODLAND PINEDROPS
Pterospora andromedea   MONOTROPACEAE

General Description This is a brownish-red plant with sticky stems up to 3 feet tall with pale-yellow flowers. Found in deep humus of coniferous forests between 2,500 to 8,500 feet, usually associated with ponderosa pine (*Pinus ponderosa*) in the Sawtooth Country. Flowers from June to August.

Ecology & Ethnobotany Some authorities prefer to call this plant a parasite rather than a saprophyte because it has no root system or fungi of its own. Instead, it draws its nutrition from mycorrhizae that are associated with adjacent plants.

Pinedrops has been used medicinally by some Native Americans. For example, a cold tea made from the pounded stems and fruits was used to treat bleeding from the lungs. As a dry powder, the plant was used as a snuff for nosebleeds. However, the plant does contain various poisonous compounds and should be avoided.

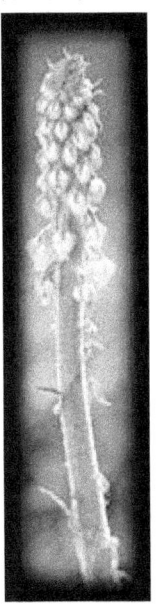

YELLOW POND-LILY
Nuphar polysepala   NYMPHAEACEAE

General Description This is a large, aquatic plant with large green, heart-shaped pads which either float or rise above the water. These plants usually obtain a spread of up to 8 feet and the rhizomes are large and the roots can grow many feet long. The flowers, are rather small when compared to the pads and are yellow and cup-shaped. They rise above the water and look rather like small "yellow balls" and appear only "half-opened" when in full bloom.

Ecology & Ethnobotany In many references it is said that the rhizomes are edible and best during the early spring and fall. They say that the starchy rootstalks can be boiled and then peeled and eaten or placed in soup or stew, or dried, ground into meal and used as flour. *THERE'S A PROBLEM WITH THAT ASSESSMENT*. My experience is that no matter how long you cook or boil or anything, it seems like it will never reach the level of tasty. *FORGET THE RHIZOME AND FOCUS ON THE SEEDS*.

The seeds can be collected and, when dry, will keep indefinitely. They can also be treated like popcorn. Simply pop them and eat or grind them into meal. The seeds can also be steamed as a dinner vegetable or cooked like oatmeal – 1-part seeds to 2 parts water.

FIREWEED
Chamerion angustifolium   **ONAGRACEAE**

General Description Previously in the genus *Epilobium*. This is robust perennial growing up to 5 feet tall. The leaves are alternate, lanceolate in shape up to 8 inches long. The edges of the leaves are more or less toothless, green colored above and pale beneath. Flowers occur in a long raceme. The 4 sepals are lavender-tinged and the petals are rose-purple in color. There are 8 stamens. Fireweed is usually found in areas disturbed by fires or moist areas in mountains. A second species, dwarf fireweed (*C. latifolium*), can be used in similar ways.

Ecology & Ethnobotany These two species were previously classified within the genus *Epilobium* (willow-herbs). Fireweed is one of the "MUST KNOW" plants when it comes to survival. Food, drink, tinder, twine, and medicine are all provided by these abundant herbs.

In general, they are survivors in landscapes that have been ravaged by manmade and natural forces (e.g., fires, clearcuts). Soil conditions do appear to affect their flavor. Many Native Americans "owned" good patches of fireweed and these were passed on to subsequent generations. The most distinctive identifying feature

of fireweed is the unique leaf venation. Unlike other plants, the veins do not terminate at the edges of the leaves, but rather join together in loops inside the outer margins.

The young shoots and leaves of fireweed may be boiled like asparagus but are better when mixed with other raw greens for a salad. The leaves, green or dry, make a good tea and are useful in settling an upset stomach. Be careful, the leaves are slightly laxative. The unopened flower buds can be used in the same manner as leaves and stems. The young fruits can also be boiled like green beans and are tasty before the seed fibers form. Mature plants tend to become tough and bitter.

The pith of the stems can also be scraped out and eaten as a snack or as a thickener for soups. If consumed in large amounts, fireweed is a gentle but effective laxative. The plant contains a relatively high content of Vitamin C and beta carotene. Raw roots are a popular food of Siberian Eskimos.

The fibrous inner bark can be used as cordage and tinder material. For use in making cordage, I found the fibers brittle. The seeds have cottonlike hairs and are great for fire starting (tinder) and insulation. Many Native Americans in the Northwest used the fluffy seed cotton as a wool substitute, mixing it with mountain goat wool or duck feathers.

CLARKIA
Clarkia   ONAGRACEAE

General Description These are annuals with brittle stems and purple or red, showy flowers. Two species in the area - pinkfairies (*C. pulchella*) and diamond clarkia (*C. rhomboidea*). They are usually found on dry slopes at the lower to middle elevations.

1. Petals 3-lobed at tip; leaves alternate ----- *C. pulchella*
1. Petals not lobed at tip; leaves subopposite ----- *C. rhomboidea*

Ecology & Ethnobotany Seeds of this genus were collected by Native Americans. After drying and parching them, the seeds were ground up into flour.

EVENING-PRIMROSE
Oenothera ONAGRACEAE

General Description There are many species in this genus which are annual, biennial, and perennial herbs. The flowers are white or yellow, often opening at night. There are 8 stamens, 4 petals, 4 sepals, and the stigma is globe-shaped to deeply four-lobed. The various species can be found in a variety of habitats up to the subalpine zone.

Ecology & Ethnobotany It has been suggested that all species would stand a trial as none are known to be poisonous. The various species are known to hybridize easily making identification at times challenging. We have cooked and eaten the young seed pods of several species and found them to have an acceptable taste. Other experts we've consulted with also suggest that many species have seeds that are edible after being parched or ground into meal.

BROOM-RAPE
Orobanche   OROBANCHACEAE

General Description All three species in the area parasitize the roots of other plants. These fleshy annual plants are nearly white to brownish or purplish in color and lack chlorophyll. The leaves are reduced to scales. Broom-rape is usually found in dry soils, associated with such genera as *Artemisia* (sagebrush) and *Eriogonum* (buckwheat). The three species are flat-

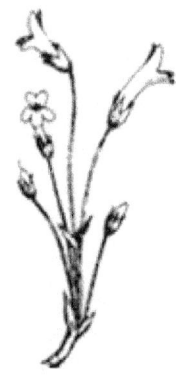

top broomrape (*O. corymbosa*), clustered broomrape (*O. fasciculata*), and one-flowered broomrape (*O. uniflora*). The following key may be useful in identifying the species.

1. Flowers sessile or on pedicels that are 12 inches long and in addition to the subtending bract have a pair of bractlets just below the calyx ----- *O. corymbosa*
1. Flowers all somewhat long-stalked and lacking small subtending bracts ----- **2**

2. Flowers no more than 2-3, usually 1, the short stems remaining underground; petals rounded; sepals narrow and slender ----- *O. uniflora*
2. Flowers more than 3; the top of stems emerging from ground; petals pointed; sepals triangular ----- *O. fasciculata*

Ecology & Ethnobotany The entire plant of broom-rape, roots and all, can be eaten raw. Being succulent plants, they answer for food and drink, and are often called sand food. We found them to be better tasting when roasted in the hot ashes of a campfire.

BROWN PEONY
Paeonia brownii   PAEONIACEAE

General Description The petals are reddish purplish with greenish-yellow borders, and the sepals are green. There are numerous stamens with golden anthers. In the center of the flower are 5 egg-shaped ovaries with no apparent styles but with a little cream-colored stigma. Leaves are light gray green. In the Sawtooth Country this species is found in sagebrush and ponderosa pine habitats.

Ecology & Ethnobotany The leaves are edible when cooked as greens. We have found it best to boil the leaves in several changes of water until the bitterness was removed.

Another method utilized by Native Americans was to pick the young leaves before the blossoms appeared in the spring, boil them, and then place in a cloth sack and weigh the sack down in the river with a stone. By allowing the

water to run through the sack overnight removed the bitterness.

TRUE FIR
Abies PINACEAE

General Description Firs are trees of cold climates with short, flat, blunt needles that grow directly from the branch. The needles are grooved above, with white stripes beneath and sometimes above. Cones mature in late summer and fall apart, leaving behind a stalk. The wood is very light, soft, and brittle, and has been used to manufacture boxes and crates. Firs are also important as pulpwood, and they make good Christmas trees since they retain their needles even when dry. The young bark often produces conspicuous pitch blisters, containing strong-smelling liquid oleoresins. The two species in the Sawtooth Country are grand fir (*A. grandis*) and subalpine fir (*A. lasiocarpa*). The following key may be useful in distinguishing them.

1. Leaves are about 1 to 2 inches long, notched at apex, and usually spreading horizontally on the branches; cones green ----- ***A. grandis***
1. Leaves mostly less than 1 inch long, turning upward, not notched at apex, and not horizontally spreading; cones purplish ----- ***A. lasiocarpa***

Ecology & Ethnobotany Subalpine fir (*A. lasiocarpa*) may be encountered at the high elevations. The inner bark can be eaten raw. The sap was chewed for pleasure and as a cure for bad breath. The boughs can be used for bedding, covering floors, and as incense.

Grand fir (*A. grandis*) may also be encountered and is also known as "stinking fir" because of an odor that may be unpleasant to some people. A brown dye can be made from the bark, and the needles were boiled to make a medicinal tea by some Native Americans.

ENGELMANN SPRUCE
Picea engelmannii   PINACEAE

General Description Engelmann spruce is a large tree up to 200 feet tall. The bark is thin, grayish, and scaly. The needles are quadrangular, stiff and sharp pointed, with white bands on all four sides. The cones develop in the tree tops and are about 1 to 3 inches long. Englemann spruce is found on cool, moist slopes up to the alpine zone usually with subalpine fir (*Abies lasiocarpa*).

Ecology & Ethnobotany The cambium or inner bark of Englemann's spruce can be eaten raw, boiled like noodles, or dried and used as a flour substitute. The inner bark was also used as a laxative. Some Native Americans ate the new growth raw, which is a good source of Vitamin C. The green spring growth on branches can be steeped in hot water for tea. Spruce beer is made from fermented needles and twigs that have been boiled with honey. The crushed needles can be rubbed on traps and skin to camouflage human scent. Although spruce seeds are very small, they are edible raw or cooked.

PINE
Pinus   PINACEAE

General Description Four species of *Pinus* can be found in the Sawtooth Country. They include whitebark pine (*P. albicaulis*), lodgepole pine (*P. contorta*), limber pine (*P. flexilis*), and ponderosa pine (*P. ponderosa*). Pines have needles in bundles of 1, 2, 3, or 5, with a membranous sheath called a bundle sheath at the base. The following key may be useful in identifying the species.

1. Leaves mostly 5 in a cluster ----- **2**
1. Leaves mostly 2 or 3 in a cluster ----- **3**

2. Cones mostly over 3 inches long, falling intact from tree; cones scales thinning toward tip; lower elevations to timberline ----- ***P. flexilis***
2. Cones mostly less than 3 inches long, seldom falling intact from tree; cones scales thickened toward tip; mostly subalpine ----- ***P. albicaulis***

3. Leaves mostly over 3 inches long, in clusters of 2 or 3; cones mostly over $2\frac{1}{2}$ inches long ----- ***P. ponderosa***
3. Leaves mostly less than $2\frac{1}{2}$ inches long, mostly in clusters of 2; cones mostly less than $2\frac{1}{2}$ inches long ----- ***P. contorta***

Ecology & Ethnobotany All pines have edible seeds. However, they are an erratic food source, yielding an abundant crop in some years and a sparse crop in others. To collect the cones, long poles were used to knock them from the branches. One of the best ways to gather seeds is to heat the green cones until they open. The seeds are best when harvested in fall or early winter when cones normally release their seeds. The nutritious seeds can then be shelled and eaten, or ground or roasted and made  into flour. Seeds may contain as much as 15% protein, 62% fat, and 18% carbohydrates, with approximately 3,000 calories per pound.

    The inner bark is also edible in an emergency. Though tedious, the tender mucilaginous layer between the bark and wood was scraped or peeled off. It was then cooked or ground into meal.

    The firm and unexpanded pollen cones can be boiled and eaten. They have a surprisingly sweet and non-pitchy taste.

    The needles of most pines can be steeped in hot water to make a satisfying tea and are a good source of Vitamin C. It also takes some practice to steep the right amount of leaves, since too much may be too strong. Additionally, the pine cleaning fluid can be extracted from boiling the needles and skimming off the oil-like substance from the surface. It may take a lot of pine needles to get a small cupful.

    Pine sap can be collected in quantity from cuts, burns, and broken branches. The collected sap is then heated and formed into balls for future use. Be careful not to expose the sap to flames as it is very flammable.

DOUGLAS-FIR
Pseudotsuga menziesii   PINACEAE

General Description This is a tall tree, 35 to 60 feet tall with drooping branches. The leaves are needle-like, blue-green, and spirally arranged on the branches, but appear to be in a flat spray because the needles are turned at the petiole base. Needles are ¾ to 1½ inches long and pointed at the tip. Cones are cylindrical, 4 to 6 inches long, with 3-fingered bracts overlapping the scales. These bracts are characteristic of the genus.

Ecology & Ethnobotany Douglas-fir is an important timber species. The wood is resinous with close, even, well-marked grains, and is of medium weight, strength, stiffness, and toughness. It is very durable, and when well-seasoned, does not warp. It is used in piles, ties, floors, and millwork, and to make a variety of items such as spear handles, spoons, fire tongs, and fishing hooks.

A tea can be made from the needles of Douglas-fir. Similar to pines, the pitch can be used as a glue. It can be used for sealing implements and caulking water containers. Medicinally, the sap provided a salve for wounds and skin irritations. The pliable roots have been used in weaving.

PLANTAIN
Plantago lanceolata & *P. major*   PLANTAGINACEAE

General Description These are characterized as short-stemmed annual or perennial herbs with basal leaves. The flowers are greenish or purplish. These two species are introduced from Europe and can be found at the lower elevations, particularly in fields and waste places.

Ecology & Ethnobotany As a food, the young leaves of both species were used fresh or cooked. They contain calcium and other minerals. One hundred grams of plantain is said to furnish as much Vitamin A as a large carrot. The older leaves may be too fibrous and bitter for use, but they are usable if one is able to remove the fibers. Seeds are tedious to collect in quantity but can be ground and used as flour substitute or extender.

CRESTED WHEATGRASS
Agropyron cristatum   POACEAE

General Description These are rhizomatous or bunch-forming perennials of mainly dry habitats. The leaves have flat or rolled blades and sheaths with free margins and usually evident auricles. The 3- to 12-flowered spikelets are sessile in a narrow, terminal spike. The 2 glumes in each spikelet are about equal in length and shorter than the lowest lemma. Both glumes and lemmas are sometimes awned from the tip.

Ecology & Ethnobotany All members of this genus are useful. The plants often grow in abundance and the seeds can be collected and ground into flour.

With some species, there may be long roots which can be been dried and ground into flour.

BENTGRASS
Agrostis   POACEAE

General Description The bentgrasses are tufted or rhizomatous annuals or perennials with sheaths that have free margins. Spikelets are stalked and borne in an open or contracted inflorescence. Each spikelet has only 1 floret. The glumes are about equally long and mostly pointed at the tip. The lemma is shorter than the glumes and has long to short hairs at the base and sometimes an awn arising from the back. The palea is very small or lacking.

Ecology & Ethnobotany Another edible group of grasses. The seeds can be made into flour

WILD OATS
Avena   POACEAE

General Description These are introduced annuals with flat leaf blades and sheaths that have free margins. The spikelets are 2- to 3-flowered and borne on flexible stalks in an open, branched inflorescence. The glumes are longer than the lowest lemma. The lemmas are leathery with a bent awn arising from near the middle.

Ecology & Ethnobotany The seeds are edible. The hairs must be singed off first and then the grains can be ground and used as flour. The seeds are approximately 15% protein and 11% fat.

BROME GRASS
Bromus POACEAE

General Description Bromes are native or introduced, annual or perennial, usually tufted grasses. The leaves have flat blades and glabrous or hairy sheaths with joined margins. The several-flowered, stalked spikelets are flattened and borne in an open to contracted inflorescence with nearly erect to drooping branches. The pointed or blunt-tipped glumes are of unequal lengths and shorter than the lowest florets. The lemmas are often awned from the tip.

Some of the introduced annual species have come to dominate disturbed grasslands throughout the semi-arid west.

Ecology & Ethnobotany The seeds of perhaps all species are edible. They can be gathered and cooked into a gruel. Tongue in cheek humor - cheatgrass (*Bromus tectorum*) - Idaho's State flower.

ANNUAL HAIRGRASS
Deschampsia danthonioides POACEAE

General Description These are tufted perennial or annual grasses with flat or rolled blades and leaf sheaths that are open along the margins. The spikelets are mostly 2-flowered and borne on spreading to erect branches of narrow or open inflorescences. Glumes are mostly of unequal length and longer than the lower floret. The lemmas are toothed at the tip and have long hairs at the base and a straight or bent awn attached about the middle of the back.

Ecology & Ethnobotany The seeds can be used in making flour

WILD RYE, SQUIRREL TAIL GRASS
Elymus   POACEAE

General Description The ryegrasses are bunch-forming or rhizomatous perennials with hollow culms and leaf sheaths with free margins. The inflorescence is a terminal spike with 2 spikelets per node. The spikelets have 2-6 florets, and the narrow glumes sometimes have a short awn and are nearly equal in length. The lemmas are rounded on the back, with or without awns.

Ecology & Ethnobotany Probably any Elymus produces seeds which can be used as food

FESCUE GRASSES
Festuca   POACEAE

General Description The members of this group of genera are annual or perennial, rhizomatous or bunch-forming grasses with hollow culms. The leaves have flat, folded, or rolled blades and sheaths with free margins. The 2- to 12- flowered spikelets are borne on erect to drooping branches of the open to contracted inflorescence. The narrow glumes have pointed tips and are usually unequal in length and shorter than the lemmas. The lemmas are rounded on the back and awnless or with a short awn-tip.

Ecology & Ethnobotany The seeds of probably all species are a food item.

FOXTAIL BARLEY
Hordeum POACEAE

General Description Our wild barleys are tufted annuals or perennials with flat leaf blades and sheaths with free margins. The mostly 1-flowered, sessile or short-stalked spikelets are borne, usually 3 per node, in a dense, terminal spike. At each node, the central spikelet has both stamens and a pistil, while the lateral spikelets only have staminate or sterile florets. The glumes are narrow and awnlike. The lemmas are rounded on the back, usually with a long awn.

Ecology & Ethnobotany Seeds were collected, ground into flour, and used as food

BLUE GRASS
Poa POACEAE

General Description These are native or introduced, annual or perennial, bunch-forming or rhizomatous grasses. The leaves have sheaths with free margins and flat, folded, or rolled blades, usually with tips shaped like the bow of a boat. The 2- to 7-flowered spikelets are borne on nearly erect to spreading or reflexed branches of the open or contracted inflorescence. The glumes are equal or unequal in length and usually much shorter than the lemmas. The lemmas are variously hairy, unawned, and often prominently 5-nerved. This is the largest genus of grasses in our area. Many of them are similar and difficult to tell apart without examination under a microscope.

Ecology & Ethnobotany Gather the seeds for food. As with many grasses, they do not remain on the plant for long, so the season for picking them is short.

INDIAN RICE GRASS
Stipa hymenoides   POACEAE

General Description The species may be known by several scientific names including *Oryzopsis hymenoides* and *Achnatherum hymenoides*. All is good as new research into genetics try to better understand how this and other plants are related to each other.

Ecology & Ethnobotany This and many other species produce edible seeds that can be eaten raw but are best when dried and ground into flour. Use the flour to make cakes and mush.

SCARLET GILIA
Ipomopsis aggregata   POLEMONIACEAE

General Description This is a biennial plant growing 1-3 feet tall. The tubular or funnel-form flowers are red, orange, pink, or white and showy. The plant is usually found on dry slopes up to the subalpine. The species has also been called skunk flower because of a faint skunk-like smell from its glandular foliage. In the Sawtooth Country, scarlet gilia grows in a variety of habitats, from desert canyons and cliffs to montane meadows, and subalpine rock fields.

Ecology & Ethnobotany The plant was used by Native Americans as a tea to treat colds, to make glue, and to treat blood troubles. In Nevada, the principal use of this plant was for the treatment of venereal diseases. The whole plant was boiled for the purpose and a solution was taken as a tea or used as a wash. The whole plant was also boiled by the Ute Indians in Utah to make glue. A blue dye can be extracted from the roots.

SLENDER PHLOX
Microsteris gracilis   POLEMONIACEAE

General Description Formerly known as *Phlox gracilis*, this small branched annual grows 4 to 8 inches high and is glandular pubescent above. Leaves are entire, and the lower ones are opposite, while the upper ones are alternate. The leaves are oblong lanceolate. Flowers are small, usually in pairs in the upper leaf axils. The corolla is salverform, with yellow tube and white to purplish pink lobes. In the Sawtooth Country slender phlox may be found in open, grassy areas below 10,000 feet. It flowers from March to August.

Ecology & Ethnobotany Slender phlox was eaten by some Native Americans as greens.

AMERICAN BISTORT
Bistorta bistortoides   POLYGONACEAE

General Description This is a slender perennial with glabrous stems. The leaves occur mostly near the base and are oblong lanceolate in shape with sheathing stipules. The upper leaves usually without petioles. Flowers occur in a thick, cylindric, pink to white terminal spike. Bistort is found in wet areas such as wet meadows and along streams between 5,000 to 10,000 feet in the Sawtooth Country.

Ecology & Ethnobotany Like *Polygonum* (knotweed and smartweed), the young greens are useful in cooking, the roots have a pleasant taste, eaten raw or cooked.

WILD BUCKWHEAT
Eriogonum POLYGONACEAE

General Description These are annual or perennial herbs; some species are woody at the base. The flowers are small and usually bright colored. The many species of *Eriogonum* can be found in various habitats at all elevations in the Sawtooth Country.

Ecology & Ethnobotany None of the species are known to be poisonous. The flowering stems can be eaten raw or cooked before they have flowered. Seeds can be collected (though tedious) and ground into flour. A tea from the root of *Eriogonum* was used to treat headaches and stomach problems. The plants are mildly astringent and were used as a gargle for sore throats.

ALPINE MOUNTAIN SORREL
Oxyria digyna   POLYGONACEAE

General Description Alpine mountain sorrel is found high in the mountains near melting snows in rocky areas. It is a low perennial with simple massed, round, thick leaves (often with red tinges) are a common and picturesque sight in subalpine and alpine rock scree. The plant has red to green flowers that are replaced by showy red/brown seed pods. Flowers in July to September.

Ecology & Ethnobotany The plant resembles a miniature rhubarb, with small rounded leaves. It has always been highly esteemed in Arctic regions as a "scurvy-grass" with an agreeable sour taste.

Perhaps one of the most refreshing plants one encounters in the high country is the alpine mountain sorrel. The new growth up to flowering time can be eaten raw, when it tastes like a mild rhubarb. The stems and leaves can be used in salads or prepared as a potherb. Some aboriginal peoples have been known to ferment mountain sorrel as a kind of sauerkraut. This is accomplished by simply letting the plant(s) sit in water for a while. This sauerkraut can then be stored for future use. The plants were also dried in the sun for traveling. Plants are high in Vitamin C and can be used to prevent and cure scurvy. Large amounts could, however, cause oxalate poisoning.

SMARTWEED, KNOTWEED
Persicaria & Polygonum   POLYGONACEAE

General Description *Polygonum* are annual or perennial herbs with stems that are more or less swollen at the nodes. The flower colors include white, greenish, or pink. They can be found in a variety of habitats up to the alpine zone. Several species of *Polygonum* have in recent years been reclassified into the *Persicaria* and *Bistorta*.

Ecology & Ethnobotany Experimentation may be the rule for *Polygonum* (and the other genera mentioned) as none of the species are known to be poisonous. They do, however, vary in degrees of palatability. Tannins are found in the plants and large amounts might cause digestive upset and possible kidney damage. In moderate quantities, however, the genus is generally regarded as safe. Based on our experiments with various species, some have peppery tasting leaves that can be used in flavoring foods. Others have starchy roots that may be eaten raw or boiled and roasted. Still others have young foliage made into good salads or potherbs.

DOCK
Rumex   POLYGONACEAE

General Description The eight species of dock in the Sawtooth Country are annual or perennial herbs. They have small flowers that are greenish and aggregated in a large terminal inflorescence. The fruits are called utricles. They can be found in many habitats in the mountains.

Ecology & Ethnobotany The young leaves of dock can be used as greens and we have found that the

flavor varies from species to species. The young leaves are best when collected before the flower stalk emerges. Also, because the leaves become watery when cooked, use very little water and don't overcook them. The older leaves may be too bitter for use. The leaves of dock are high in Vitamin C and contain more Vitamin A than carrots. Native Americans ground dock seeds and used the meal to make breads. However, removing the papery seed cover involves a lot of work, and depending on the species, is probably more work than it is worth. The distinctive sour taste of these plants is due to oxalic acid. As with other species that contain oxalic acid, docks should be used in small portions as they can cause calcium deficiency.

RED MAIDS
Calandrinia menziesii   PORTULACACEAE

General Description Formerly known as *Calandrinia ciliata*. This is a low glabrous annual plant growing up to 16 inches tall. The leaves are alternate, entire, linear to oblanceolate in shape and somewhat fleshy. The flowers are a magenta or red-violet color and there are 2 sepals and 5 petals. This is a fairly common species found growing in early spring on grassy slopes within the Sawtooth Country.

Ecology & Ethnobotany The seeds of all *Calandrinia* species are edible. After gathering and winnowing, the seeds should be parched with coals, pulverized, and then pressed into cakes for eating. The roots were also eaten, and the young leaves and stems were eaten raw or cooked.

BITTERROOT
Lewisia rediviva   PORTULACACEAE

General Description Bitterroots are indigenous to the western parts of the U.S. and can be found clinging precariously to rocky ledges among boulders, on rock-strewn slopes, damp gravely places, alpine meadows, and in near desert conditions where rainfall is seasonal and unpredictable. A few of the species are limited in their distribution and considered rare.

Ecology & Ethnobotany Although all species may be edible, *L. rediviva* is the species that has been used extensively. This species was an important food item for many Native Americans. The root is remarkably large and thick for a small plant and contains nutritious farinaceous matter that is much prized. The roots are dug up in spring before flowering. Once dug, the root is peeled promptly and the small red "heart" (embryo of next year's growth) is removed to reduce the roots bitter flavor. It is then steamed, boiled, or pit-cooked and eaten. The root can also be dried and will keep for a long time. The bitterness of the root varies and cooking is said to improve the flavor. The root boiled to a jellylike consistency will be pink in color. The pounded root was chewed for a sore throat.

Though some still collect it today, bitterroot is considered a rare plant in many areas. Overgrazing and trampling by range livestock and habitat destruction from agricultural encroachment seem to have been a major impact on many *Lewisia* populations. Remember, digging the roots destroys the plant. Programs to maintain and enhance habitat for the plant are recommended.

SPRING BEAUTY, MINER'S LETTUCE
Claytonia & Montia   PORTULACACEAE

General Description These are two closely related genera that may be best treated here together.

Claytonia - These are as perennial succulent herbs arising from deep corms or fleshy taproots. The leaves are opposite and the flowers white or pink in color.

Montia - This genus is comprised of slightly succulent annual and perennial herbs. The flowers have two persistent sepals and five white or pinkish petals. Most *Montia* species grow in moist or seasonally wet areas that are partially to fully shaded. Formerly, several of the claytonias were classified as montias.

Ecology & Ethnobotany *Claytonia* - Often called Indian potato, wild potato, or mountain potato, the small corms can be eaten raw, boiled, or roasted. For many Native Americans, spring beauty was an important "root vegetable." When collecting, keep only the largest corms and replant the others. At first, many find the corms distasteful, as they do take a little getting used to. The corms are high in starch and, when cooked, taste like potatoes. Boil or bake the corm for 30 minutes. Most species are not plentiful, so be conservative in your endeavor. They can also be dried on strings for long-term storage. The rosettes can also be eaten raw

or cooked and are high in Vitamins A and C. They are better when mixed with other salad plants.

Miner's lettuce (*C. perfoliata*) has stems and leaves that can be eaten raw or boiled like spinach. The roots are also edible raw or boiled. Some Native Americans picked miner's lettuce and placed it near the nests of red ants. The ants were allowed to crawl over the leaves and were then shaken off. The residue left on the leaves by the ants had an acerbic flavor.

All species of *Montia* have stems and leaves that can be eaten raw or boiled like spinach. The roots are also edible raw or boiled.

COMMON PURSLANE
Portulaca oleracea   PORTULACACEAE

General Description This is a small succulent annual herb found in disturbed habitats at the lower elevations. **One caution** – DO NOT COLLECT the plant along busy roads where a variety of toxins may have made their way into the soil (e.g., lead from leaded gasoline use in the past). It has been used as a food for more than 2,000 years in India and Persia. In Europe, it is grown as a garden vegetable. The genus name may be derived from *portula* meaning "little gate," referring to the lid on the capsule.

Ecology & Ethnobotany This plant have been used for centuries. The foliage is good in salads and excellent when cooked. It has been pickled, used to thicken soups, or dried or frozen for storage. Seeds are used in the form of flour.

PONDWEED
Potamogeton   POTAMOGETONACEAE

General Description Pondweeds are perennial aquatic plants with extensive, slender rhizomes and simple branched stems that often root at the nodes. The leaves have stipules that clasp the stem. The lower leaves are alternate and submerged, and the upper often floating leaves are wider and opposite. The small, greenish flowers are clustered in a spike that arises from the upper leaf axils.

Ecology & Ethnobotany Probably all pondweeds have starchy, edible rhizomes, but species with larger rootstocks are preferred for gathering. One species (*P. diversifolius*) was an important source of strong fibers which were rolled into cordage to make carrying nets, rabbit-trap nets, and other items.

SHOOTING STAR
Primula   PRIMULACEAE

General Description Previously included in the genus *Dodecatheon*. They perennial herbs and all leaves are basal and form a loose rosette. The flowers are located at the end of a stalk with narrow, reflexed rose colored petals. The species habitats range from grassland to shrubland, meadows, and riparian habitats

up to the alpine zone. At least 3 species occur in the Sawtooth Country.

Ecology & Ethnobotany Since none of the species are listed anywhere as poisonous, it is likely that all the species are edible. It is usually the texture that discourages people from using the plants. We have found that the leaves of many species have a good flavor when eaten raw.

COLUMBIAN MONKSHOOD
Aconitum columbianum   RANUNCULACEAE

General Description This is a perennial herb with palmately divided or lobed leaves. Flowers are usually deep blue or purple but may also be pale to white. Usually found in moist, densely shaded places often with streamside vegetation up to timberline.

Ecology & Ethnobotany Some species of monkshood have been a source of drugs, as a pain killer or as a sedative for nervous disorders. However, *all parts of the plant are poisonous and should be considered dangerous* if ingested.

BANEBERRY
Actaea rubra   RANUNCULACEAE

General Description This is a perennial herb with fibrous roots. The leaves have long petioles and are 2-3 times divided into sharply toothed, lance-shaped segments. The small flowers are white and borne in a branched, congested, hemispheric inflorescence. The fruits are shiny red or white. Baneberry is common in moist, montane forests and riparian areas, usually with some partial shade.

Ecology & Ethnobotany The *entire plant, especially the berries, is poisonous*. The plant is sometimes confused with *Osmorhiza chilensis* (western sweetroot) which often shares the same habitat. However, unlike baneberry, sweetroot has a strong licorice-like odor.

For those of you having been scouts remember the adage "berries white poisonous site." Keep in mind baneberry produces both white and red berries and that in no way are the red berries edible. *I have met scout leaders with absolutely no experience in the outdoors and botany assume only that white berries are poisonous and that the red one are edible.*

COLUMBINE
Aquilegia formosa   RANUNCULACEAE

General Description This perennial herb with large divided leaves has red and yellow flowers that are nodding. Each of the 5 petals extends backward between the sepals forming a hollow spur. Its is

generally found in moist habitats, subalpine meadows and on slopes.

Ecology & Ethnobotany The flowers of columbine are edible and have a sweet taste but grow bitter with age. They can be added to salads in small amounts. **Warning** The seeds can be fatal if eaten and most parts of the columbine contain cyanogenic glycosides. Any therapeutic use of columbine is strongly discouraged.

☠ WHITE MARSH-MARIGOLD
Caltha leptosepala   RANUNCULACEAE

General Description The leafless flowering stem of this plant has one or two flowers with 5-15 white or blue-tinged, petal-like sepals. There are no petals. The leaves are dark green and basal. In the Sawtooth Country, marsh-marigold can be found in marshes and wet meadows to above timberline. Marsh-marigolds bloom close to receding snowbanks. Marsh-marigold is not related to the cultivated marigold a member of the Sunflower Family.

Ecology & Ethnobotany The young leaves can be used as a potherb and the spaghetti-like roots can be dug up during the winter and boiled as a pasta substitute. Though the plant is poisonous when raw (the plant contains the poisonous glucoside, protoanemonin), cooking appears to destroy the poison. It also contains the deadly glucoside hellebrin, which breaks down with boiling. In his book "*Stalking the Helpful Herbs*," Euell Gibbons reports a drop of juice squeezed from the fresh leaves is caustic and will remove warts.

☠ BUTTERCUP
Ranunculus   RANUNCULACEAE

General Description Buttercups are either perennials or occasionally annual herbs with simple to compound leaves. The flowers are solitary or borne in a small inflorescence. The 5 petals are normally yellow or white and have a nectar gland at the base. They can be found in many different habitats from the lower elevations to the alpine zone.

Ecology & Ethnobotany All species are more or less poisonous when raw. The leaves and stems should be boiled in several changes of water to remove the poisonous compounds. The volatile toxin is also rendered harmless by drying. The seeds can be parched and ground into meal for bread or pinole. The roots can also be boiled and eaten and were an important part of some Native American diets. A yellow dye can be obtained by crushing and washing the flowers. All species have corrosive juice which is very painful if rubbed into the eyes.

SNOWBRUSH
Ceanothus velutinus   RHAMNACEAE

General Description Snowbrush is easy to identify by its shiny, often sticky, evergreen leaves with 3 main veins. Its small, creamy white flowers are borne in pyramidal clusters.

Ecology & Ethnobotany The genus has been long recognized as a substitute for commercial black tea and the leaves and flowers could be used to make tea. The seeds can also be used as food. Many species contain saponin which gives the flowers and fruits their soap-like qualities. The flowers when crushed and rubbed in water, will produce a light lather for purposes of washing oneself. The long, flexible shoots were used in basketry. The red roots yield a red dye.

SERVICEBERRY
Amelanchier   ROSACEAE

General Description These are shrubs or small trees with simple leaves that are serrate on the terminal half. The white flowers have 5 petals and 5 reflexed sepals, and many stamens. The ovary is inferior and the fruit a pome (apple-like). In the Sawtooth Country three species may be encountered and all can be used in similar ways.

Ecology & Ethnobotany All species produce edible pomes that ripen in late spring and the summer. They were a considered to be a major food for many

Native Americans. In fact, some Native Americans intentionally moved their camps to locations where they could be more easily harvested. The pomes may be eaten raw, cooked, or dried. After drying, the pomes can be pounded into loaves or cakes. These in turn may be eaten after softening a piece in water or placing them in soups or stews. Prepared this way, the pomes could be kept for several years. Additionally, the dried pomes could be incorporated into pemmican. The wood can be used for arrows, digging sticks, and other useful items.

CURL-LEAF MOUNTAIN-MAHOGANY
Cercocarpus ledifolius　　ROSACEAE

General Description This is a shrub or small tree with hard wood and entire, lanceolate-shaped leaves. The flowers are not showy, but instead are sweet with nectar. The fruit is quite characteristic and is an achene that ends with a long terminal style that is covered with shiny hairs at maturity. The shrubs glisten in the sun from the mass of silvery fruits, each one a "tailed fruit" as indicated by the generic name. It is fairly common on dry mountain slopes between 4,000 to 10,000 feet.

Ecology & Ethnobotany Native Americans used the wood for spears, arrow shafts, and digging sticks. The inner brown bark produced a red-purple dye, as did the roots. The leaves and seeds of curl-leaf mountain-mahogany contain cyanogenic glycosides and should be considered toxic.

HAWTHORNE
Crataegus   ROSACEAE

General Description In general, they are large deciduous shrubs or small trees with thorns. The leaves are toothed or lobed and the white flowers are borne in an open inflorescence. The fruits are small pomes, borne in tremendous quantity, and remaining on the tree all winter. *Crataegus* is a large and varied genus containing many species that readily hybridize.

Ecology & Ethnobotany All species produce edible, albeit mealy fruits which may be eaten raw or cooked in small amounts, or dried and mixed into pemmican. A diet high in hawthorne pomes or drinking hawthorne tea is said to reduce weight. The pomes contain a non-toxic heart stimulant and should not be eaten in large amounts or without admixture. The pomes also contain Vitamin C. The thorns have many practical uses such as prongs or rakes, lances for blisters, piercing ears, and as fish hooks.

WILD STRAWBERRY
Fragaria   ROSACEAE

General Description Two species, *F. vesca* (woodland strawberry) and *F. virginiana* (Virginia strawberry), occur in the Sawtooth Country. These white-flowered perennial herbs are produced from rootstocks and have long runners that root at the nodes. The leaves are clustered at the base of the stem and are divided into three egg-shaped, coarsely toothed leaflets. They can be found in moist, humus-rich, well-drained soils of open forest and forest margins up the subalpine zone.

1. Apical tooth of leaflets greater than those on either side; leaves yellow-green, upper surface bulged between the veins ----- *F. vesca*
1. Apical tooth of leaflets smaller than those on either side; leaves blue-green ----- *F. virginiana*

Ecology & Ethnobotany Strawberries do not keep well and should be dried for future use if not eaten soon after being picked. Tea made from the green or dried leaves is said to tone up one's appetite. Externally, the leaf tea can also be used as an antiseptic wash for eczema and wounds and as a gargle for sore throat and mouth ulcers. The plants do contain substantial amounts of Vitamins A and C, and sulphur, calcium, potassium, and iron. To remove tartar, rub the berries on your teeth and let the juice sit for a few minutes. Afterwards, brush your teeth thoroughly with baking soda and water.

OCEANSPRAY
Holodiscus discolor   ROSACEAE

General Description This is a deciduous shrub growing up to 10 feet tall with grayish-red bark. The egg-shaped leaves are shallowly lobed with toothed margins. The creamy-white colored flowers are small and borne in a diffusely branched inflorescence. This shrub

grows at in the woods or fairly moist areas between 4,500 to 11,000 feet. The plant is found in areas prone to wildfire, and it is often the first green shoot to spring up in an area recovering from a burn.

Ecology & Ethnobotany The small dry berries can be used as food, eaten raw or cooked. The hard wood can be used for digging sticks.

SILVERWEED
Potentilla anserina   ROSACEAE

General Description There are many species of *Potentilla* In the Sawtooth Country. In general, they are perennial, biennial, or annual herbs, and this species is a shrub. The flowers are yellow, white, or, in one case, purple. They can be encountered in various habitat types at all elevations. Silverweed is generally found moist to sort-of alkaline places between 4,000 to 8,000 feet.

Ecology & Ethnobotany The large fleshy, older roots of silverweed can be boiled or roasted and added to soups and stews. Prepared this way they are quite tasty and have a nutty or a parsnip-like texture, but more woody tasting. They were a staple among many Native Americans. Today they are seldom harvested, but greatly enjoyed by those who still use them. Silverweed is high in tannins and can be used to tan leather.

BITTER CHERRY, CHOKECHERRY
Prunus   ROSACEAE

General Description These are shrubs or trees with simple leaves. Many species have a pair of warty glands present at the tops of the petioles or at the bases of the leaf blades. Flowers are pink to white and rather showy. Fruits are drupes with 1 stone and typically embedded in the fleshy pulp. In our area you usually encounter bitter cherry (*P. emarginata*) and common chokecherry (*P. virginiana*). The following key may be useful in distinguishing them.

1. Flowers numerous, in unbranched, long, narrow inflorescences ----- ***P. virginiana***
1. Flowers in hemispheric or flat-topped inflorescences ----- ***P. emarginata***

Ecology & Ethnobotany In general, the fruits of all species are sour or bitter when raw, but after cooking the sourness disappears. Native Americans dried the berries whole or in cakes for use in winter. When needed, the dried fruits were soaked in water and then eaten.

To make cakes, the ripe fruits are usually ground up, pits and all, and dried in the sun. When needed, the cakes, or portions thereof, can be soaked in water, mixed with flour and sugar and made into a sauce or gravy. This sauce was eagerly traded among some Native Americans. The only difficulty we've found in preparing cakes in this manner is that the pits do not grind down nicely into a fine material, leaving larger chunks that could have resulted in broken teeth.

Other uses of the berries included their incorporation into pemmican. They can also be used in making jelly, but because *Prunus* are low in natural pectin, it is advisable to add pectin.

The leaves of the species contain toxic amounts of cyanide as do the seeds (pits). Cyanide is highly volatile and the pits can be rendered safe by long-term drying, by boiling in several changes of water, or by dry roasting. Do not eat them in significant amounts even then unless you mix them with larger quantities of other foods. Prunus shoots, peeled and split, were used in basketry. The wood was used for various implements, such as digging sticks, arrows, and arrow fore shafts.

BITTERBRUSH
Purshia tridentata   ROSACEAE

General Description This fragrant shrub grows up to 8 feet tall. Leaves are deeply 3-cleft into linear lobes, glandular above and hairy below. The leaf margins are rolled inwards. Flowers are pale yellow to white and the fruit is a pubescent oblong achene. Bitterbrush grows on dry slopes and canyons of the mountains between 3,000 and 9,000 feet. Flowers from April to June.

Ecology & Ethnobotany The ripe seed coat produces a violet dye. Old *Purshia* stumps produce shredding bark that can be peeled off, worked to soften, and used as toilet paper or material with which to start a fire by friction.

WILD ROSE
Rosa ROSACEAE

General Description These are shrubs with prickles with leaves that are pinnately divided into 3-11 leaflets. The large, red to pink flowers are borne singly or a few together. The fruits, called hips, are orange, red, or purplish and urn-shaped.

Ecology & Ethnobotany The hips are edible raw, stewed, candied, or made into preserves. They are high in Vitamin C and also contain Vitamins E, B, and K, beta-carotene, calcium, iron, and phosphorus. There are many other edible parts, besides the fruit. Young Rosa shoots in spring make an excellent potherb, and the roots and stems can be used to make a tea. The petals may be used in salads. The peeled spring shoots can also be nibbled upon. Almost all parts of the plant have been made into a wash or dressing for cuts or sores to coagulate blood. One of the more common methods is to sprinkle fine shavings of de-barked stems into a washed wound. The petals can be used as a dressing. A poultice of leaves can be used to relieve insect stings. In addition, the young leaves can be washed, cut into small pieces, and dried for a hot tea.

BLACKBERRY, RASPBERRY, THIMBLEBERRY
Rubus ROSACEAE

General Description These species are deciduous shrubs with arching or trailing stems covered with bristles and prickles. The flowers have white petals and the fruit is a coherent cluster of small, 1-seeded drupes (raspberries, blackberries, dewberries, cloudberries, marionberries).

Ecology & Ethnobotany All species produce edible fruits. Flowers can be added to salads and can be nibbled upon when hiking. The fresh or dried leaves can be steeped for a tea, alone or in herbal blends. Do not use the wilted or molded foliage, as it may be toxic. The young shoots cut just above ground can be peeled and eaten raw or cooked.

MOUNTAIN-ASH
Sorbus scopulina   ROSACEAE

General Description This species is a shrub up to 13 feet tall with reddish-brown branches and buds and young twigs that are finely white- or grayish-hairy. The leaves have 11-17 shiny and nearly glabrous, broadly lance- shaped leaflets with finely toothed margins. The narrow appendages at the base of the petioles fall early in the season. The white, oval petals are about ¼ inch long, and the berries are orange to red and shiny but without a waxy coating. It can be found in moist meadows and forest openings at the middle elevations.

Ecology & Ethnobotany The fruits may be eaten raw, cooked, or dried. They are high in Vitamin A and C, and carbohydrates. Unripe berries are very bitter and somewhat unpalatable. The fruits, which are pomes, are commonly processed into jams and jellies. They have high pectin content and jell readily. As a coffee substitute, grind the dried, roasted seeds. The berry juice can be used as a gargle for sore throat and as an antiseptic wash for cuts. <u>sorbitol</u>, the sugar in the fruit of *Sorbus* is being used commercially for sweetening candies, toothpaste, and other products.

SPIRAEA, MEADOWSWEET
Spiraea   ROSACEAE

General Description Three species of meadowsweet can be found in the Sawtooth Country. They are small shrubs with deciduous leaves with white to pink flowers that are densely clustered in showy, flat-topped to spike-like inflorescences. The various species can be found in brushy, open slopes and to moist habitats up to timberline.

Ecology & Ethnobotany *Spiraea* is a source of methyl salicylate, similar to the active ingredient in aspirin. Native Americans brewed a tea from the stem, leaves, and flowers of some species to use as a pain reliever. The plants are astringent and a poultice made from the leaves and bark was used to treat ulcers, burns, and tumors. The roots were also peeled and boiled until soft, mashed and used as a poultice for burns. The wiry, branching twigs can be used to make broom-like implements for collecting tubers.

BEDSTRAW, CLEAVERS
Galium aparine   RUBIACEAE

General Description Despite their small flowers, the various species of *Galium* are unmistakable. They are annual or perennial herbs with 4-angled stems and whorled leaves. The small, 4-parted flowers are white or greenish and the fruits are smooth or bristly hairy. They can be found in various habitats from the low to higher elevations.

Ecology & Ethnobotany None of the species of *Galium* are known to be poisonous. Although *G. aparine* is

the most commonly used species, it is believed that all other species can be used similarly. The very young leaves and stems can be used as a potherb. The small hairs on the stems make the plant difficult to swallow raw, boiling or steaming, however, does soften them up. If the stems are too fibrous, use only the leaves. Slow roasted until dark brown and ground, the ripe fruit can be used as a coffee substitute.

ASPEN, COTTONWOOD
Populus tremuloides & *P. trichocarpa*   SALICACEAE

General Description Cottonwoods are trees with sticky, resinous leaf buds, and deciduous leaves. Older trees of some species have gray, rough bark; young bark is smooth and whitish. The flowers are borne in catkins that appear before the leaves. Cottonwoods are usually associated with streams.

Ecology & Ethnobotany The catkins may be eaten raw or boiled in stews and are a source of Vitamin C. The inner bark can also be eaten as a spring tonic, or dried and ground into a flour substitute or extender. The fresh or dried plant can be used in poultices for muscle aches, sprains, or swollen joints. The primary action of *Populus* is that of an analgesic, used topically and internally. It contains varying amounts of populin and salicin, compounds related to early forms of aspirin. The leaves and bark are most effective parts for tea and aid in diarrhea problems. The wood makes for an excellent bow and drill fire set. Cottonwoods are considered to be botanical indicators of water.

WILLOWS
Salix   SALICACEAE

General Description Many species of willow can be found in the Sawtooth Country. They are mostly shrubs with numerous stems. Flowers are in catkins that appear before, with, or after the leaves. Willows generally grow along streams or other moist habitats

Ecology & Ethnobotany The young shoots and leaves can be eaten raw. The bitter inner bark can also be eaten raw, although it is better dried and ground into flour substitute or extender. The plant contains salicin which is similar to aspirin and useful as a substitute. Any part of the willow can be used to produce a tea for use as an aspirin replacement for headache and body pain. The highest concentrations of salicin, however, are found in the inner bark. Because it is not nearly as strong as aspirin, you may have to drink quite a bit of it.

The leaves have astringent properties that are effective when placed on wounds and cuts. Bark was chewed as a toothache remedy. Bark, leaves, twigs, and roots produced medicinal teas, powders, washes, and poultices to relieve pain, swelling, infection, bleeding, and many other ailments. Willows, like the cottonwoods, are botanical indicators of water. The branches of many willow species are very flexible and make them very useful for traps, arrow shafts, and other needs, such as basketry. Fiber from bark was used for cordage, nets, and clothing.

BASTARD TOADFLAX
Comandra umbellata   SANTALACEAE

General Description Bastard toadflax is a partially parasitic perennial herb with a waxy surface and a rather woody base. The leaves are linear and the flowers are bell-shaped. The fruit is a 1-seeded, berry-like drupe. It is common and widespread in shrublands up to the subalpine zone. The roots are blue when cut.

Ecology & Ethnobotany The mature, brown, urn-shaped fruit of bastard toadflax may be eaten raw, and is best when slightly green. They were popular with Native Americans because of their sweet taste. The berries, however, are rarely found in sufficient quantities for more than a pleasant tidbit. Consuming too many berries may cause nausea.

ALUMROOT
Heuchera   SAXIFRAGACEAE

General Description In general, *Heuchera* species are perennial herbs with basal leaves. Flowers are small, saucer- to bell-shaped, and greenish, white or pinkish in color. Their various habitats include moist soils and rocky areas up to the alpine zone. The three species that occur in the area include roundleaf alumroot (*H. cylindrica*), Bridger Mountain alumroot (*H. flabellifolia*), and gooseberryleaf alumroot (*H. grossulariifolia*).

1. Calyx deeply campanulate (bell-shaped) to urn-shaped ----- *H. cylindrica*
1. Calyx short-campanulate to saucer-shaped ----- **2**

2. Calyx campanulate, yellowish, sepals erect ----- *H. grossulariifolia*
2. Calyx widely top-shaped to saucer-shaped, sepals spreading ----- *H. flabellifolia*

Ecology & Ethnobotany The leaves of all species are edible, although they are not choice. They have a sour taste because of the high tannin content. Therefore, the leaves should be boiled or steamed. Since they are rather tough, we found them to be more palatable if chopped and added to soups or salads.

Heuchera is said to be one of the strongest astringents due to their high tannin content, as much as 20 percent their weight in tannins. Tannins tend to shrink swollen, moist tissues. As such, alumroots are also gastrointestinal irritants and have been known to cause kidney and liver failure. Ingestion of the plant should be in moderation. Otherwise, the pounded, dried roots of many species have been used as a poultice that stops bleeding and promotes healing when applied to cuts and sores. The raw root, eaten in small amounts, has been used as a cure for diarrhea. A tea from the roots can also be used as a gargle for sore throats. The powdered roots have been used as an antiseptic.

SAXIFRAGE
Micranthes & Saxifraga   SAXIFRAGACEAE

General Description There are many species of saxifrage in the Sawtooth Country. In recent years, research has reclassified many species of the genus *Saxifraga* into *Micranthes*. In general, they are perennial herbs with basal, alternate, or opposite leaves. The flowers are broadly bell-shaped. The many species occur in various habitats from the foothills to alpine. The generic name is from Latin *saxum* meaning rock and *frangere*, meaning to break, and alludes to many species rocky habitat.

Ecology & Ethnobotany The genera as a whole are regarded as a safe group of plants. The leaves can be used fresh or in stews and are high in Vitamins A and C. In China, some species were used in the treatment of nausea and ear infections. In our area, there is little documentation regarding medicinal uses.

PAINTBRUSH
Castilleja   SCROPHULARIACEAE

General Description This is a large genus found primarily in western North America that contains many species. The genus is easily recognized, but many species are notoriously difficult to identify. They are perennials with deeply lobed to entire leaves. The flowers are subtended by colorful leaf-like bracts. Some paintbrushes are partial root parasites found in various habitats up to the alpine zone.

Ecology & Ethnobotany Many, if not all of the species have flowers and bracts that can be eaten raw. The seeds of some species were gathered, winnowed, dried, and stored for winter use. In winter they were parched, pounded and eaten dry. The plants, however, absorb selenium from the soil and so should be taken in moderation. Symptoms in humans of selenium poisoning will vary with the amount and form ingested, but may include difficulty in breathing, excessive urine production, loss of appetite, mental depression, a weak and rapid pulse, blurry vision, digestive upset, and eventually coma and death.

YELLOW MONKEYFLOWER
Mimulus guttatus SCROPHULARIACEAE

General Description Many species of monkeyflower can be found in the Sawtooth Country. They are annual or perennial herbs with opposite leaves. The flowers flare at the mouth to form 5 lobes, 2 which form the upper lip and 3 lobes that form the lower.

Yellow monkeyflower is a perennial growing up to 40 inches tall, and has hollow stems. Leaves are

opposite, oval to rounded in shape, finely toothed, and 3/8 to 3 inches long. The lower leaves are long-petioled, while the upper leaves are sessile and slightly fused together at their bases. Flowers are on pedicels and are ¾ to 2½ inches long. Calyx is bell-shaped, inflated, and is tinged with red and 5-angled, 5/8 to 1-inch long. The yellow corolla is 2-lipped, with red spots, 5/8 to 1½ inches long. Yellow monkeyflower is common along streams and in wet places and can be found throughout the area below 10,000 feet. Flowers from March to August.

Ecology & Ethnobotany The young stems and leaves of *M. guttatus* have been used as salad greens. Sometimes, leaves were burned and the ash used a salt. Weedon (1996) indicates that the young herbage of

Mimulus species may be eaten in salads, and that they grow bitter with age, but remain edible. For example, the leaves of *M. primuloides* were eaten by the Maidu Indians in California and the young plants of *M. moschatus* were boiled and eaten by some Native Californians.

WOOLLY MULLEIN
Verbascum thapsus   SCROPHULARIACEAE

General Description This is a coarse plant with stems up to 6 feet tall. The basal leaves are broadly lance-shaped, entire-margined or shallowly toothed and up to 16 inches long. The stem leaves are smaller, and the upper ones clasp the stems, forming low wings below the point of attachment. The whole plant is densely covered with long soft hairs. Flowers are borne in a densely congested inflorescence. The saucer-shaped flowers are white or yellow. The upper 3 stamens are densely hairy. The species can be found in disturbed places up to the subalpine zone.

Ecology & Ethnobotany The leaves of mullein are said to be edible when eaten in small quantities and cooked. Because of their woolly texture, however, we have found the plants to be undesirable. The leaves can also be used as

poultices applied locally to hemorrhoids, sunburn, and inflammations. The dried stalks are ideal for use as hand-drills to start fires. The flowers and leaves produce yellow dye; as a toilet paper substitute, the large fresh leaves are choice.

SPEEDWELL
Veronica   SCROPHULARIACEAE

General Description Several species of speedwell occur in the Sawtooth Country. They are annual or perennial herbs with opposite, alternate, or, rarely, whorled leaves. The small, blue, pink, lilac, or white flowers are saucer-shaped and 4-lobed, the upper lobe being the largest. The mature fruit is often necessary for identification. Most species are found in wet soils or shallow waters from low to high elevations.

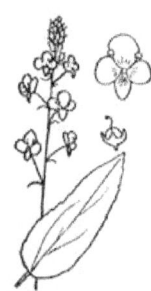

Ecology & Ethnobotany The leaves and stems of all species, when collected during the spring and early summer, can be eaten like watercress, added to salads, or prepared as potherbs. The taste of the various species ranges from spicy to bitter to bland depending on personal taste. The plants also contain moderate amounts of Vitamin C and were once used to prevent scurvy. The leaves and stems can also be steeped as a tea. Care should be taken to avoid plants growing in polluted waters.

MATTRIMONY VINE
Lycium barbarum   SOLANACEAE

General Description This is an introduced species. In general, *Lycium* are shrubby plants with entire to minutely toothed leaves that sometimes grow in bundles. The purple to greenish-purple flowers are borne solitary, or in small clusters in the leaf axils. The roundish berries are fleshy to dry, depending on the species.

Ecology & Ethnobotany The fruit of matrimony vine (and other *Lycium* species) are edible raw or cooked and has a mild sweet liquorice flavor. Only the fully ripe fruits should be eaten, unripe berries could be poisonous. As a food, dried wolfberries are traditionally cooked before consumption.

Their taste has an accent of tomato and is similar to that of dates, dried cranberries or raisins; though drier, more tart, and less sweet and with an herbal scent. Dried wolfberries are also used frequently in raw food diets.

BUR-REED
Sparganium angustifolium & *S. natans*   TYPHACEAE

General Description Bur-reeds are aquatic perennials with unbranched, erect or floating stems. The leaves are linear and sheath the stem. The flowers are borne in dense, round clusters. Bur-reeds can be found in shallow waters of marshes, ponds, and slow-moving steams.

Ecology & Ethnobotany The bulbous bases of the stems and tubers of *S. angustifolium* (narrowleaf burreed) can be used as food in much the same way as cattails (*Typha*) and bulrushes (*Scirpus*): dried and pounded into flour.

BROADLEAF CATTAIL
Typha latifolia   TYPHACEAE

General Description Cattails are found over much of North America. A cattail plant produces a basal cluster of narrow, ribbon-like leaves that are several feet long and stand almost vertically. The upright stem is unbranched, not quite as long as the leaves, and it bears a long, dense, brown spike at the upper end. The spike may vary from 4 to more than 12 inches long. Its upper part bears stamens intermixed with long hairs, each stamen constituting a flower, while its lower part bears pistillate flowers, each flower consisting only of an ovary with an abundance of dark hairs at its base. A second species known as narrowleaf cattail (*T. angustifolia*) may also be encountered. It can be used similarly. The following key may be useful in identifying the species.

1. Staminate and pistillate parts of spike usually contiguous or almost; stigmas oblanceolate to obovate in shape; leaves over $\frac{1}{2}$ inch wide ----- ***T. latifolia***
1b. Staminate and pistillate parts of spike separated by at least $\frac{1}{4}$ inch; stigmas linear; leaves less than $\frac{1}{2}$ inch wide ----- ***T. angustifolium***

Ecology & Ethnobotany Virtually every part of these plants has a use, from food to fiber. In fact, Native Americans and wilderness adventurers consider the cattail the "supermarket of the swamps." Although both cattail species have edible rhizomes, the rhizomes should never be raw since they may cause vomiting. The rhizomes should be boiled or roasted or dried and then ground into meal or flour.

When pulling up the rhizome, you may notice newly emerging buds. These can be scrubbed, peeled, and eaten raw or boiled. The swollen joint between bud and rhizome is also starchy. Peel it, then roast or boil for a potato-like vegetable. Like the rhizomes, this part should not be eaten raw. The young green shoots can be peeled of their green outer layer and eaten raw or cooked. It is always good to boil them in a couple of changes of water if there is any bitterness. The peeled core can also be sliced and added to salads.

Useful fibers can be derived from cattails. Fibers in stems can be loosened by soaking plant material in water for several days. The silky fluff on the seeds is buoyant and water repellent and makes a good insulator, especially in boots. The silk can be used for stuffing items from pillows to down vests. It can also be used for tinder. The fuzz will explode into flame with a spark from a flint and steel set. Leaves can be woven to make mats, sandals, baskets, etc. The stems provide a good coil foundation for baskets. Additionally, the stalks have been used as arrows and hand drills. A toothbrush can be fashioned from the fuzzy stem with the flowers removed.

NETLEAF HACKBERRY
Celtis laevigata var. *reticulata* ULMACEAE

General Description This is a small tree or shrub with leaves that are ovate to lanceolate in shape, with entire to serrate edges. The fruit is a drupe, and the plants can usually be found growing along streams or on dry canyon slopes at the lower elevations.

Ecology & Ethnobotany The small orange, red, or yellow fruits are edible raw and have a sweet taste to them. The entire fruits can also be dried and then ground into a flour.

STINGING NETTLE
Urtica dioica URTICACEAE

General Description This is an annual or perennial herb with stinging hairs. The flowers are numerous, small, and clustered on drooping branches at the base of the leaves. Stinging nettle can be found along roadsides, streams, in moist areas and waste places in the low to middle elevations. Stinging nettle is an indicator of good soil conditions.

Ecology & Ethnobotany The young stems and leaves of stinging nettle are edible after boiling and are very delicious as a spinach substitute. Boiling the leaves destroys the formic acid found in the hairs. The leaves are high in vitamins A, C, and D, the latter of which is rare in plants. The roots are also edible after they have been roasted.

The older stems become fibrous, which reduces their edible qualities, but allows them to be used to produce strong cordage. The older leaves also contain cystoliths that can irritate the kidneys. A yellow dye may be obtained by boiling the roots.

VALERIAN
Valeriana   VALERIANACEAE

General Description These are perennial herbs with aromatic (actually ill-smelling) roots. The stem leaves are opposite and pinnately compound and the flowers have three stamens. They can be found in open forests and meadows to timberline and above.

Ecology & Ethnobotany Valerian, in general, are considered best known for their calming qualities. They have been used for more than a hundred years as a remedy for anxiety, muscle tension, and insomnia. The plants contain valepotriates, which is a known herbal calmative, antispasmodic, and nerve tonic, and is used for hypochondria, nervous headaches, irritability, and insomnia. Research has confirmed that teas, tinctures, or extracts of this plant are a central nervous system depressant and a sedative for agitation.

The roots and leaves of *V. edulis* (edible valerian) can be collected, steamed for a day or two to remove the disagreeable odor, and then used in soups as a potato substitute. However, the taste does get a little getting used to and may remain somewhat unpalatable. We found that the steamed roots are better if dried, ground into flour, and then added to other flours. The other species could probably also be used in an emergency, but they do not have the large taproot of *V. edulis*.

VIOLET
Viola VIOLACEAE

General Description Violets are low-growing, perennial or annual herbs. The leaves are spade-shaped and basal. The flowers occur singly on the ends of stems and have five petals. There are 2 upper and 2 lateral petals, and 1 lower petal that is prolonged into a nectar holding pouch at the base of the flower. Most species also have small, self-fertilizing flowers that do not open. In the Satooth Country, violets can be found in meadows and open forests from the foothills to above timberline.

Ecology & Ethnobotany The leaves, buds, and flowers of possibly all species are edible raw or cooked, with some being more palatable than others; the leaves make a good tea. Adding the leaves to soups make them thicker. Violets are high in Vitamin C and beta-carotene.

Collect the plants by leaving the roots intact. Since many species reproduce vegetatively, you will probably not inhibit next year's growth significantly. Many naturalists indicate that all violets are safe for consumption, but there are some experts that insist some yellow species may be somewhat purgative. All species do have a tendency to be slightly laxative, so proceed slowly. The flowers have also been candied or made into jellies and jams.

INDEX

A

Abies · 162
Acer · 37
Achillea · 55
Achnatherum · 172
Aconitum · 129, 184
Actaea · 146, 185
ADOXACEAE · 100
Agastache · 134
AGAVACEAE · 142
Agoseris · 5, 56
Agropyron · 167
Agrostis · 168
ALDER · 82
Alisma · 38
ALISMATACEAE · 38
ALLIACEAE · 140
Allium · 140, 147
Alnus · 82
ALUMROOT · 205
Alyssum · 86
ALYSSUM · 86
AMARANTH · 39
AMARANTHACEAE · 39
Amaranthus · 39
Amelanchier · 118, 190
ANACARDIACEAE · 40
Anaphalis · 57

Antennaria · 58
Anthemis · 76
APIACEAE · 41, 43, 44, 45, 46, 47, 48, 49, 50, 51
APOCYNACEAE · 52, 53
Apocynum · 52
Aquilegia · 185
Arabis · 87
ARACEAE · 139
Arctic Alpine Forget Me Not · 85
Arctium · 59
Arctostaphylos · 119
Arnica · 60
ARNICA · 60
Artemisia · 61, 160
Asclepias · 53
ASH · 200
ASPEN · 202
ASTERACEAE · 55, 56, 57, 58, 59, 60, 61, 63, 65, 66, 67, 68, 69, 70, 71, 72, 73, 74, 75, 76, 77, 78, 79, 80
Atriplex · 106
Avena · 168

B

Balsamorhiza · 63, 80
BALSAMROOT · 63
BANEBERRY · 185
Barbarea · 88
BARLEY · 171
Bassia · 107
Bearberry · 120
BEARGRASS · 150
BEAVERTAIL · 97
BEDSTRAW · 201
BEEPLANT · 99
BENTGRASS · 168
BERBERIDACEAE · 81
Berberis · 81
Betula · 84
BETULACEAE · 82, 84
BIRCH · 84
BISTORT · 174
Bistorta · 174, 177
BITTERBRUSH · 197
BITTERCRESS · 90
Bitterroot · 79
BITTERROOT · 179
BLACKBERRY · 199
BLAZINGSTAR · 152
BLUE FLAG · 133
BLUE GRASS · 171
BLUEBELL · 98
BLUEBELLS · 85
BLUEBERRY · 122
Boechera · 87

BORAGINACEAE · 85, 131, 133
BOUNCING BET · 103
BRACKENFERN · 117
Brasenia · 96
Brassica · 88, 89
BRASSICACEAE · 86, 87, 88, 89, 90, 91, 92, 93, 94, 95
Brodiaea · 147
BROME GRASS · 169
Bromus · 169
BROOM-RAPE · 160
BUCKBEAN · 154
BUCKWHEAT · 175
BUFFALOBERRY · 117
BUGLEWEED · 136
BULRUSH · 116
BURDOCK · 59
BUR-REED · 211
BUTTERCUP · 187

C

CABOMBACEAE · 96
CACTACEAE · 97
Calandrinia · 179
Calochortus · 141, 147
Caltha · 186
CAMAS · 142, 147
Camassia · 142, 147
Camelina · 89
Campanula · 98

CAMPANULACEAE · 98
CAPPARACEAE · 99
CAPRIFOLIACEAE · 100, 101
Capsella · 89
Cardamine · 90
Carex · 115
CARYOPHYLLACEAE · 102, 103, 104
Castilleja · 207
CATNIP · 138
CATTAIL · 212
Ceanothus · 189
CELASTRACEAE · 105
Celtis · 214
Centaurea · 15
Cerastium · 102
CERATOPHYLLACEAE · 105
Ceratophyllum · 105
Cercocarpus · 191
Chaenactis · 65
Chamerion · 156
chamomile · 76
CHEESEWEED · 153
CHENOPODIACEAE · 106, 107, 108, 109, 110
Chenopodium · 5, 107
CHERRY · 195
CHICKWEED · 104
CHICORY · 67
CHOKECHERRY · 195
Chrysothamnus · 66, 69

Cichorium · 67
Cicuta · 23, 41, 42, 46
Cirsium · 67, 110
Clarkia · 158
CLARKIA · 158
Claytonia · 5, 181
CLEAVERS · 201
Cleome · 99
CLOVER · 125
COLUMBINE · 185
Comandra · 204
COMANDRACEAE · 204
Common Kochia · 107
Conium · 43
Conyza · 68
COON'S TAIL · 105
CORNACEAE · 111
Cornus · 111
Corypantha · 97
COTTONWOOD · 202
COWPARSNIP · 45
CRASSULACEAE · 112
Crataegus · 192
Crepis · 69
Croton · 123
CUPRESSACEAE · 113
CURRANT · 129
Cymopterus · 44
CYMOPTERUS · 44
CYPERACEAE · 115, 116
Cyperus · 115

D

DANDELION · 78
DENNSTAEDTIACEAE · 117
Dentaria · 90
Deschampsia · 169
Descurainia · 91
DOCK · 178
Dodecatheon · 183
DOGBANE · 52
DOGWOOD · 111
DUCKWEED · 139
Dustymaiden · 65
Dysphania · 108

E

ELAEAGNACEAE · 117
ELDERBERRY · 100
ELKWEED · 127
Elymus · 170
Epilobium · 156, 157
EQUISETACEAE · 118
Equisetum · 118, 130
Eremocarpus · 123
ERICACEAE · 119, 121, 122, 155
Ericameria · 69
Erigeron · 68, 70
Eriogonum · 160, 175
Eriophyllum · 71
Eritrichium · 85
Erodium · 127, 128
Erysimum · 91
Erythronium · 143
Escobaria · 97
EUPHORBIACEAE · 123
EVENING-PRIMROSE · 159
EVERLASTING · 58

F

Fabaceae · 123
FABACEAE · 124, 125, 126
False Flax · 89
FESCUE GRASSES · 170
Festuca · 170
FIR · 162, 166
FIREWEED · 156
FLATSEDGE · 115
FLAX · 152
Fleabane · 70
Floerkea · 151
Fragaria · 192
Frasera · 127
FRINGE-POD · 95
Fritillaria · 144, 147
FRITILLARY · 144

G

Galium · 201
Gaultheria · 120
GENTIANACEAE · 127
GERANIACEAE · 127, 128
Geranium · 128, 129
GERANIUM · 128
GILIA · 172
GLACIER LILY · 143
Glycyrrhiza · 123
GOAT'S BEARD · 80
GOLDENBUSH · 69
GOLDENROD · 77
GOLDENWEED · 79
GOOSEBERRY · 129
GOOSEFOOT · 107
Grindelia · 72
GROSSULARIACEAE · 129
Gumweed · 72

H

HACKBERRY · 214
HAIRGRASS · 169
Haplopappus · 69
HAREBELL · 98
HAWKSBEARD · 69
HAWTHORNE · 192
Helianthus · 72, 81
HENBIT · 135

Heracleum · 45
Heuchera · 205
Hippuris · 130
Holodiscus · 194
HONEYSUCKLE · 100
Hordeum · 171
HOREHOUND · 136
HORSE MINT · 134
HORSETAIL · 118
HORSEWEED · 68
HUCKLEBERRY · 122
HYDRANGEACEAE · 131
Hydrophyllum · 131
HYPERICACEAE · 111
Hypericum · 111

I

Indian hemp · 52
INDIAN POTATO · 48
INDIAN RICE GRASS · 172
Ipomopsis · 172
IRIDACEAE · 133
Iris · 133

J

JERUSALEM OAK GOOSEFOOT · 108
JUNIPER · 113
Juniperus · 113

K

Kalmia · 121, 122
KINNIKINNICK · 119
KNOTWEED · 177
Kochia · 107

L

Lactuca · 73
LAMIACEAE · 134, 135, 136, 137, 138, 139
Lamium · 135
Lathyrus · 126
Layia · 74
LAYIA · 74
Lemna · 139
Lepidium · 92
LETTUCE · 73, 181
Lewisia · 179, 180
LICORICE · 123
LICORICE ROOT · 46
Ligusticum · 46, 47
LILIACEAE · 141, 143, 144, 145, 146
LILY · 149, 155
LIMNANTHACEAE · 151
LINACEAE · 152
Linum · 152
LOASACEAE · 152
Lomatium · 47
LOMATIUM · 47
Lonicera · 100
LOVAGE · 46
LUNGWORT · 85
LUPINE · 124
Lupinus · 124
Lycium · 211
Lycopus · 136

M

Madia · 75
MADWORT · 86
Maianthemum · 145, 146
MALLOW · 153
Malva · 5, 153
MALVACEAE · 153
MANZANITA · 119
MAPLE · 37
MARE'S-TAIL · 130
MARIPOSA LILY · 141
Marrubium · 136
MARSH-MARIGOLD · 186
Matricaria · 75
MATTRIMONY VINE · 211
MEADOWSWEET · 201
MELANTHIACEAE · 147, 148, 149, 150
Melilotus · 125
Mentha · 137

Mentzelia · 152, 153
MENYANTHACEAE ·
  154
Menyanthes · 154
MERMAIDWEED · 151
Mertensia · 85
Micranthes · 206
Microseris · 76
Microsteris · 173
MILKWEED · 53
Mimulus · 208, 209
MINT · 137
MOCKORANGE · 131
MONKEYFLOWER ·
  208
MONKSHOOD · 184
Monolepis · 109
Montia · 181, 182
MONTIACEAE · 179,
  181
MONUMENT PLANT ·
  127
MOUNTAIN
  DANDELION · 56
MOUNTAIN LAUREL ·
  121
MOUNTAIN LOVER ·
  105
MOUNTAIN-
  MAHOGANY · 191
MOUSE EARS · 102
MUGWORT · 61
MULLEIN · 209
MUSTARD · 88

N

Nasturtium · 5, 93
Nepeta · 138
NETTLE · 214
Nodding Silverpuff ·
  76
Nuphar · 155
NYMPHAEACEAE · 155

O

OATS · 168
OCEANSPRAY · 194
Oenothera · 159
ONAGRACEAE · 156,
  158, 159
ONION · 140
Opuntia · 97
OREGON-GRAPE · 81
OROBANCHACEAE ·
  160, 207
Orobanche · 160
Orogenia · 48
Oryzopsis · 172
Osmorhiza · 49, 185
Oxyria · 176

P

Pachistima · 105
Paeonia · 161
PAEONIACEAE · 161

PAINTBRUSH · 207
PARSLEY · 47
PARSNIP · 51
Paxistima · 105
PEONY · 161
PEPPERGRASS · 92
PEPPERWEED · 92
Perideridia · 50
Persicaria · 177
Phacelia · 133
PHACELIA · 133
Philadelphus · 131
PHLOX · 173
PHRYMACEAE · 208
Picea · 163
PIGWEED · 39
PINACEAE · 162, 163, 164, 166
Pincushion · 65
PINE · 164
PINEAPPLE WEED · 75
PINEDROPS · 155
Pinus · 155, 164
PLANTAGINACEAE · 167, 210
Plantago · 167
PLANTAIN · 167
Poa · 171
POACEAE · 167, 168, 169, 170, 171, 172
POISON HEMLOCK · 43
POLEMONIACEAE · 172, 173

POLYGONACEAE · 174, 175, 176, 177, 178
Polygonum · 174, 177
PONDWEED · 183
Populus · 202
Portulaca · 182
PORTULACACEAE · 182
Potamogeton · 183
POTAMOGETONACEAE · 183
Potentilla · 194
POVERTYWEED · 109
PRICKLY PEAR · 97
Primula · 183
PRIMULACEAE · 183
Prince's Plume · 95
Prunella · 139
Prunus · 195, 196
Pseudostellaria · 102, 103
Pseudotsuga · 166
Pteridium · 117
Pterospora · 155
Purshia · 197
PURSLANE · 182
PUSSYTOES · 58

R

RABBITBRUSH · 66, 69

RANUNCULACEAE ·
  184, 185, 186, 187
Ranunculus · 187
RASPBERRY · 199
RED MAIDS · 179
RHAMNACEAE · 189
Rhodiola · 112
Rhus · 40
Ribes · 129
ROCKCRESS · 87
Rorippa · 94
Rosa · 198
ROSACEAE · 190, 191,
  192, 194, 195, 197,
  198, 199, 200, 201
ROSE · 198
RUBIACEAE · 201
Rubus · 118, 199
Rumex · 178
RUSCACEAE · 145

S

SAGEBRUSH · 61
SALAL · 120
SALICACEAE · 202,
  203
Salix · 203
SALSIFY · 80
Salsola · 110
SALTBUSH · 106
Sambucus · 100
SAND SPURRY · 103

SAPINDACEAE · 37
Saponaria · 103
Saxifraga · 206
SAXIFRAGACEAE ·
  205, 206
SAXIFRAGE · 206
Schoenoplectus · 116
Scirpus · 116, 212
SCROPHULARIACEAE
  · 209
SEDGE · 115
Sedum · 112, 113
SELF-HEAL · 139
SERVICEBERRY · 190
SHEPHERD'S PURSE ·
  89
Shepherdia · 117
SHOOTING STAR ·
  183
SILVERWEED · 194
Sisymbrium · 94
Sium · 51
SMARTWEED · 177
Snow Drops · 48
Snowberry · 120
SNOWBERRY · 101
SOLANACEAE · 211
Solidago · 77
Sonchus · 77
Sorbus · 200
SORREL · 176
Sparganium · 211
SPEARMINT · 137
SPEEDWELL · 210

Spergularia · 103
spicywintergreen · 120
SPICYWINTERGREEN
  · 120
Spiderflower · 99
SPINYSTAR · 97
Spiraea · 201
SPIRAEA · 201
SPRING BEAUTY · 181
SPRUCE · 163
SQUIRREL TAIL
  GRASS · 170
ST. JOHN'S WORT ·
  111
Stanleya · 95
STARWORT · 102
Stellaria · 102, 103,
  104
Stenotus · 79
Stipa · 172
STONECROP · 112
STORKS-BILL · 127
STRAWBERRY · 192
Streptopus · 5, 146
SUMAC · 40
SUNFLOWER · 72
SWEETCICELY · 49
SWEETCLOVER · 125
SWEETROOT · 49
Symphoricarpos · 101
Syringa · 131

T

TANSYMUSTARD · 91
Taraxacum · 56, 78
Tarragon · 62
TARWEED · 75
THEMIDACEAE · 148
THIMBLEBERRY · 199
THISTLE · 67, 77, 110
Thysanocarpus · 95
TOADFLAX · 204
Tonestus · 79
TOOTHWORT · 90
Toxicoscordion · 143,
  147
Tragopogon · 80
Trifolium · 5, 125
Trillium · 148
TRILLIUM · 148
Triteleia · 148, 149
TRITELEIA · 148
TULE · 116
TUMBLEMUSTARD ·
  94
TURKEY-MULLEIN ·
  123
TWISTED-STALK ·
  146
Typha · 212
TYPHACEAE · 211, 212

U

ULMACEAE · 214
Urtica · 5, 214
URTICACEAE · 214

V

Vaccinium · 122
VALERIAN · 215
Valeriana · 215
VALERIANACEAE · 215
Veratrum · 149, 150
Verbascum · 209
Veronica · 210
VETCH · 126
Vicia · 126
Viola · 216
VIOLACEAE · 216
VIOLET · 216

W

WALLFLOWER · 91
WATER HEMLOCK · 41
WATER PLANTAIN · 38
WATERCRESS · 93
WATERLEAF · 131
WATERSHIELD · 96
Western Pearly-everlasting · 57
WHEATGRASS · 167, 170
WILD RYE · 170
WILLOWS · 203
WINTERCRESS · 88
WORMWOOD · 61
Wyethia · 63, 80
WYETHIA · 80

X

Xerophyllum · 150

Y

YAMPAH · 50
YARROW · 55, 71
YELLOWCRESS · 94

Z

Zigadenus · 24, 147

www.ingramcontent.com/pod-product-compliance
Lightning Source LLC
Chambersburg PA
CBHW071413170526
45165CB00001B/266